HARVARD HISTORICAL STUDIES

Published under the direction of the
Department of History at the charge
of the Henry Warren Torrey Fund and
with a subsidy from La Trobe University

Volume XCV

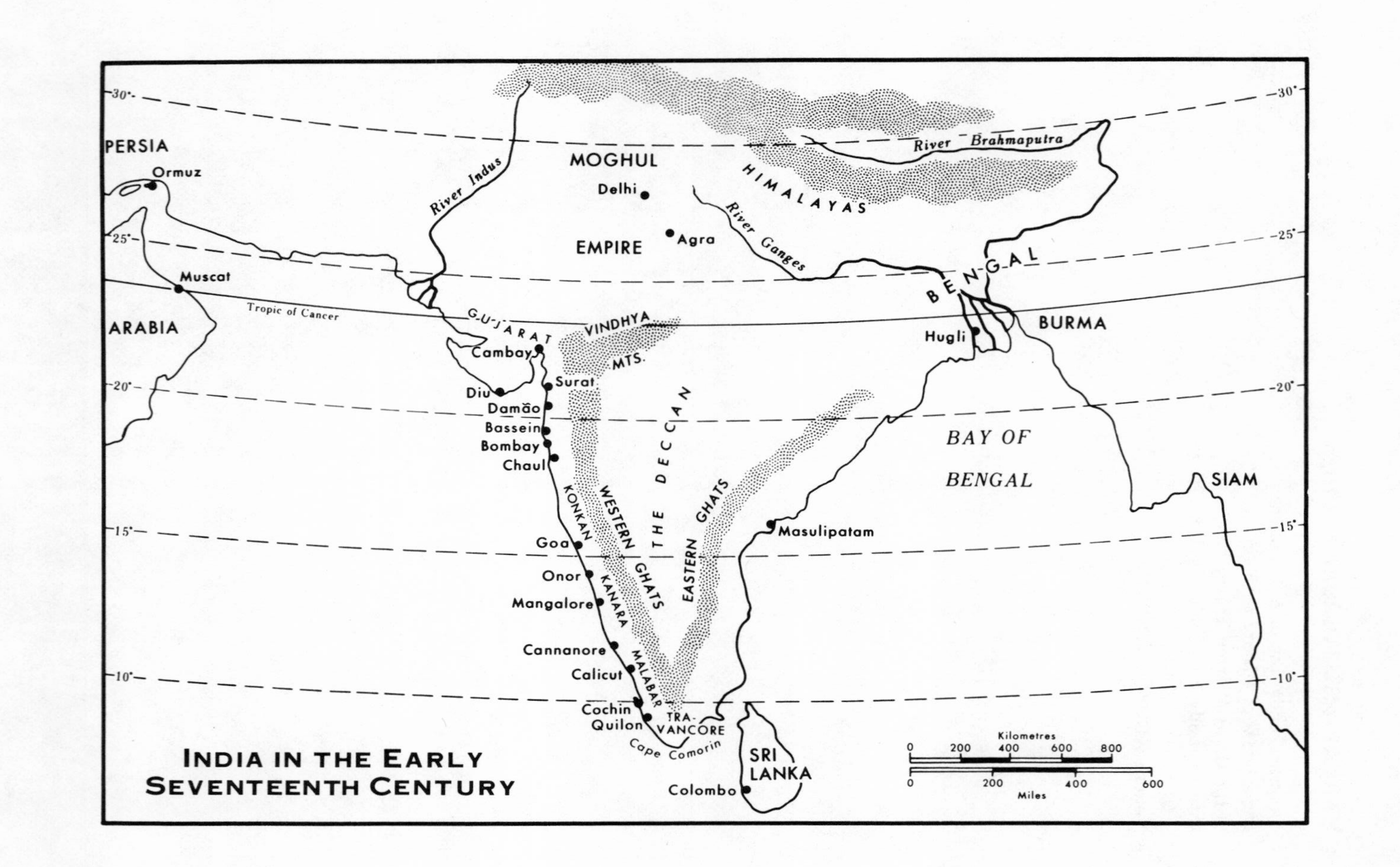

INDIA IN THE EARLY SEVENTEENTH CENTURY

Twilight of the Pepper Empire

Portuguese Trade in Southwest India in the Early Seventeenth Century

A. R. Disney

Harvard University Press
Cambridge, Massachusetts
and London, England
1978

Printed in the United States of America
Library of Congress Cataloging in Publication Data
Disney, Anthony R 1938-
Twilight of the pepper empire.

(Harvard historical studies; 95)
Bibliography: p.
Includes index.
1. Portugal—Commerce—India—History.
2. India—Commerce—Portugal—History. 3. Pepper
(Spice) I. Title. II. Series.
HF3698.I4D57 1978 382'.09469'054 77-17376
ISBN 0-674-91429-5

Preface

Vitorino Magalhães Godinho wrote in 1958 that, with the partial exception of Brazil, "all, or almost all" still remained to be done in the economic and social history of the Portuguese expansion. This statement failed to take account of a small number of serious studies then available on the Portuguese in Asia, such as the early works of C.R. Boxer on China and Japan; but as a broad generalization it was, at the time, substantially valid. During the past two decades, however, significant works dealing with certain parts of the empire have been published. On the economic history of the Portuguese in Asia there is now M.A.P. Meilink-Roelofsz's survey of trade in the Indonesian archipelago in the sixteenth and early seventeenth centuries (1962), and Boxer's account of the life and business activities of the seventeenth-century Portuguese merchant, Francisco Vieira de Figueiredo (1967). Magalhães Godinho's massive study of the economic structure and development of the early Portuguese empire appeared initially in serial form between 1965 and 1968, and in a somewhat modified French version in 1969; Niels Steensgaard's analysis of the fall of Ormuz, and of the Portuguese struggle against the Dutch and English East India Companies, was published in 1973.

Despite such progress, the history of the Portuguese commercial empire in the Indian subcontinent itself has remained curiously neglected. Only Michael N. Pearson's valuable monograph, *Merchants and Rulers in Gujarat,* with its analysis of Portuguese and Gujarati trade, seriously studies any region or aspect of Portuguese India. The present book is an attempt to break some new ground on this little-understood subject. Geographically the book concentrates on Kanara and Malabar, and—insofar as it is related to these regions—on the viceregal capital of Goa. Kanara and Malabar in southwest India were selected because, as the principal sources of pepper—easily the most important commodity sought by the Portuguese crown in Asia—they appeared to have critical significance. The chronological limits of the book are broadly the Hapsburg period, with occasional incursions into the years immediately before and after; but the viceroyalty of the Count of Linhares at Goa between 1628 and 1636 has been selected for special study. These years constituted a decisive phase in the crisis that overtook Portuguese Asia in the early seventeenth century, a prolonged crisis essentially economic and organizational, though its more dramatic manifestations were military. It was during Linhares' viceroyalty that an attempt was finally made to modernize the Lisbon-Goa trade by forming a Portuguese India Company, modelled on English and Dutch precedents. This company concentrated overwhelmingly on the pepper trades of Kanara and Malabar; its antecedents, the difficulties of its conception and birth, its trading history and ignominious failure are here traced in detail. The collapse of the company in 1633 was followed by the signing two years later of what proved a lasting truce with the English East India Company based at Surat. These developments, which spelled the effective abandonment by the Portuguese of their monopoly claims to the Cape trade route, and their reluctant recognition that northern European traders were in Asia to stay, form the climax of the book, although some brief attention is also given to the effects that followed.

For the sake of convenience, certain conventions have been adopted throughout. To avoid confusion, the Hapsburg kings are styled by their better-known Spanish titles, although strictly speaking Philip II is Philip I of Portugal, and so on. Portuguese proper names have generally been modernized, rather than left in the often inconsistent seventeenth-century spellings. The names of institutions, offices, titles, etc. have been translated into English except where this appeared clumsy or unconventional, in which cases Portuguese forms have been retained, but with English equivalents when first introduced.

The gathering of material for this study involved visits to Portugal, Spain and England in 1968 and 1973, to Brazil in 1969 and to India in 1971. I am greatly indebted to Harvard and La Trobe Universities for making these visits possible. In all the countries concerned, librarians and archivists gave indispensable assistance, and I am grateful in particular to the directors and staffs of the Arquivo Nacional do Tôrre do Tombo, the Biblioteca Nacional, the Biblioteca da Ajuda and the Arquivo Histórico Ultramarino in Lisbon, the Biblioteca Pública e Arquivo Distrital at Évora, and the Archivo General de Simancas at Simancas, Spain. In Brazil the Biblioteca Nacional and the Instituto Histórico e Geográfico Brasileiro of Rio de Janeiro were equally helpful, while in Panjim, India, the consideration of Dr. Guné and the courteous assistance of Mr. Ghantkar at the Archive of the Indies were particularly appreciated. I owe a special debt to Dr. Lawrence W. Towner and his staff at the Newberry Library, Chicago, where I spent a pleasant and productive year, through the generosity of the Gulbenkian Foundation of America and of the Newberry itself. To Mr. Garside and the Library of King's College, London, I am indebted for a photocopy of most of the Codex Lynch, which has been a key source for much of this book. Many hours of conversation with Mr. Valter Lopes in Lisbon have helped me improve my understanding of the Portuguese world of the early seventeenth century. From colleagues at the Newberry and in the Department of History at La Trobe, and from Professor Francis M. Rogers in the Department of Romance Languages and Literatures at Harvard, I have received valuable criticism and advice. In particular, however, Professor John H. Parry, through his guidance, encouragement and support, has earned my lasting gratitude.

Anthony Disney

Contents

1 / The Portuguese in Kanara and Malabar

Dominating central India, the vast plateau of the Deccan extends in a huge triangle from the Vindhya Mountains in the north to Cape Comorin in the south. Along the two coastal rims of the plateau run narrow, fertile lowlands, separated from the interior by the low ranges of the Western and the Eastern Ghats which culminate in the Nilgiri and Anaimalai hills in the extreme south.[1]

Of the two coastal ranges that fringing the western side of India is the higher. Worn and rugged, its gneiss and granite peaks thrust up to an average height of some 3,000 feet, and attain almost 9,000 feet in the highest points of the Anaimalais and Nilgiris. From these heights the Ghats slope gently away on the inland side, but are steep and occasionally precipitous on the western approaches. Here deep valleys crease the mountains, and through them cascading rivers and streams break out to the lush lowlands that fringe the Arabian Sea.

From north to south along this coast lie the regions of Gujarat, the Konkan, Kanara and Malabar. Their coastal lowlands vary in width from about twenty to fifty miles, being narrowest in the Konkan and north Kanara, and broadest in south Kanara and Malabar. Rich in rice and coconuts, especially in the south, they have long been densely peopled, while on the hill-slopes grows much luxuriant tropical forest which even today covers one-fifth of the state of Kerala.*

A hot wet climate makes such vegetation possible. From early June to mid-September the western coast of India receives the heavy rain-bearing currents of the southwest monsoon. Torrential downpours, broken by bursts of brilliant sunshine while the mist and clouds swirl down from the mountain valleys, characterize this period, during which 90 per cent of the annual precipitation occurs. On the coastal plains the level of rainfall approximates 100 inches a year, but the average increases to 300 inches on the crests of the Western Ghats.

In Malabar the force of the monsoon seas and winds has built up sandbanks along the coast, which have helped form many picturesque

*Kerala State (created in 1956) comprises roughly the old region of Malabar, and parts of southern Kanara.

and quiet backwaters well suited to small coastal and river craft. Farther north in Kanara and the Konkan the sandbanks fade, and a firm shoreline extends in a long curve towards Bombay. Except for Bombay itself, Marmagao and Cochin, there are no good deepwater harbors, although numerous sheltered creeks and roadsteads provide for smaller vessels. Along the whole coast from Gujarat to Travancore the orientation has traditionally been towards the sea and seaborne trade, and the social flavor cosmopolitan.

However, there was and is a marked difference in linguistic patterns between the north and south. The northern sectors of the coast, especially Gujarat and the Konkan and to a lesser extent north Kanara, were deeply affected by successive waves of Aryan invaders beginning about 1500 B.C., and in these regions Marathi—an Indo-European language—came to predominate. By contrast, south Kanara and Malabar, more remote from the main land routes into India and more sharply cut off from the heart of the subcontinent by the Western Ghats, remained relatively immune to Aryan influence. They continued predominantly Dravidian, their inhabitants speaking Kanarese, Tulu and Malayalam.

When Vasco da Gama reached Calicut in 1498, India south of the Deccan was dominated by the Hindu empire of Vijayanagar, soon to attain the height of its power under its ablest and greatest ruler, Krishna Deva Raya (1509-1529). Two generations later Vijayanagar was crushingly defeated at Talikot by an alliance of Moslem sultanates from the north (1565), and broke up into a number of autonomous Hindu princedoms, of which Tanjore, Madura and Ikkeri were the most important. Kanara, like the rest of southern India, remained politically fragmented—until, during the first quarter of the seventeenth century, it fell under the control of the nayaks of Ikkeri.

The emergence of Ikkeri as the de facto paramount power in Kanara was largely due to Venkatapa Nayak who undertook a series of expansionist wars in the region between 1602 and his death in 1629. He was followed as nayak by Vira Bhadra, a strong-willed grandson, who to enforce his succession eliminated a rival claimant, crushed a Portuguese-supported rebellion of his subject princelings, and staved off a threatened invasion from Bijapur. Both Venkatapa and Vira Bhadra were nominal vassals of the titular kings of Vijayanagar who by this time were effective rulers of only the small princedom of Chandragiri.

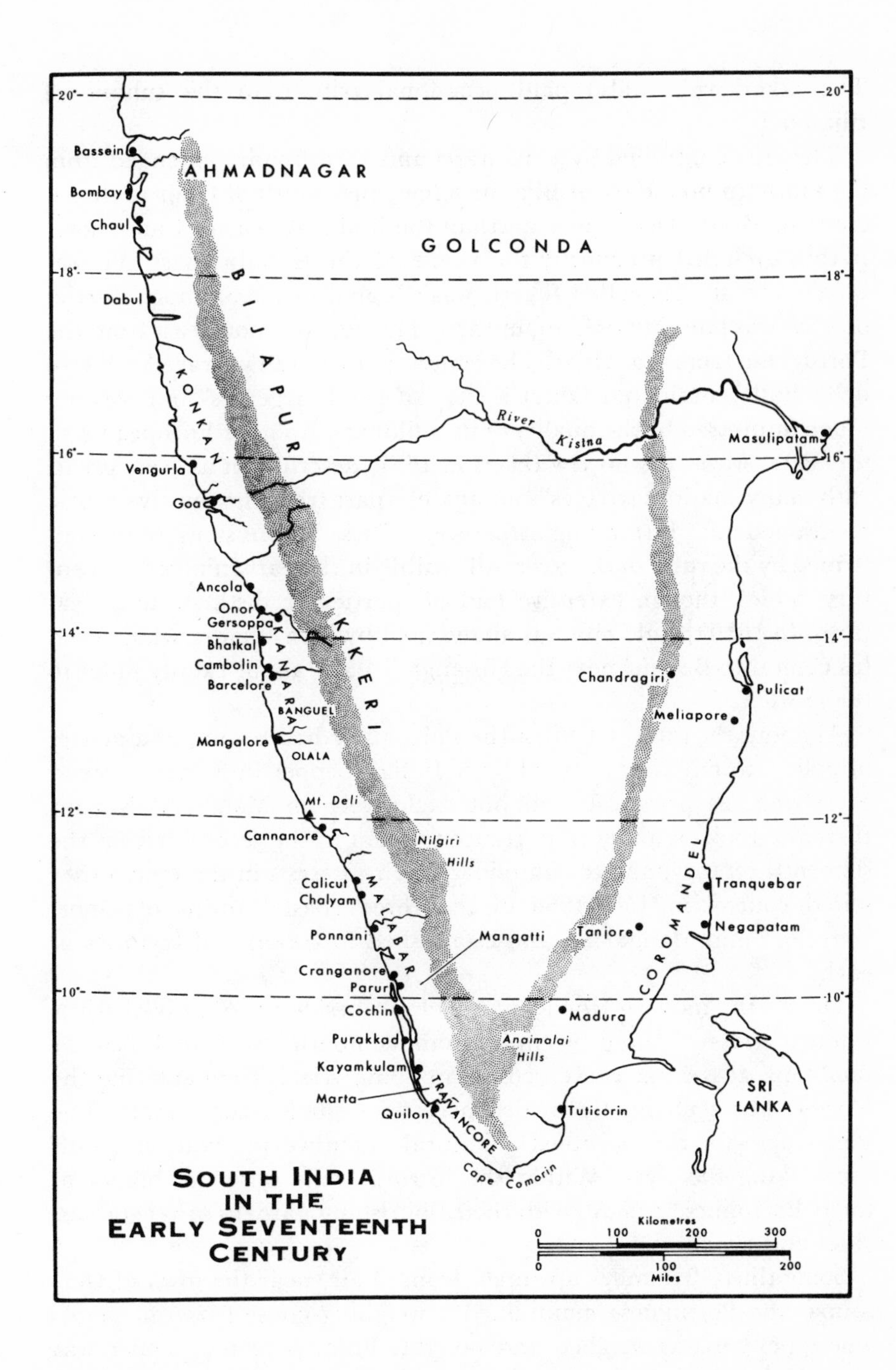
20°
18°
16°
14°
12°
10°
Bassein
Bombay
Chaul
Dabul
AHMADNAGAR
GOLCONDA
BIJAPUR
KONKAN
River Kistna
Masulipatam
Vengurla
Goa
Ancola
Onor
Gersoppa
Bhatkal
Cambolin
Barcelore
KANARA
IKKERI
BANGUEL
Mangalore
OLALA
Mt. Deli
Cannanore
Nilgiri Hills
Calicut
Chalyam
MALABAR
Ponnani
Mangatti
Cranganore
Parur
Cochin
Purakkad
Kayamkulam
Marta
Quilon
TRAVANCORE
Cape Comorin
Anaimalai Hills
Madura
Tuticorin
Tanjore
COROMANDEL
Chandragiri
Pulicat
Meliapore
Tranquebar
Negapatam
SRI LANKA
SOUTH INDIA IN THE EARLY SEVENTEENTH CENTURY
Kilometres
0 100 200 300
0 100 200
Miles

The Ikkeri rulers also paid occasional tribute to the sultans of Bijapur.[2]

The area controlled by Venkatapa and Vira Bhadra extended from the southern boundary of Bijapur a few miles north of Onor to the vicinity of Mount Deli a little north of the Malabar port of Cannanore. It thus included practically the whole of the Kanara coast. Venkatapa's capital, also called Ikkeri, was situated in the Shimoga district of what was later Mysore, about three days' journey northeast from the Portuguese factory at Onor. The contemporary Italian traveller Pietro della Valle considered Ikkeri a city "of good largeness" but was not much impressed by the quality of its buildings. Another European visitor, Peter Mundy, who saw Ikkeri in 1637, described it as "very greatt with many spatious streetes" but again, apart from the massive tanks, mentioned no outstanding structures. These impressions were confirmed by the ruins of the town still visible in the early nineteenth century, which though extensive had no particular distinguishing features. In 1639 the old city was abandoned by Vira Bhadra who moved his capital to Bednur near the Hosangadi Pass, about twenty miles to the south.[3]

At about the time of Talikot the Portuguese had begun, as a matter of policy, to purchase some of their Indian pepper in Kanara instead of relying, as previously, all but exclusively on Malabar.[4] Kanara therefore took on a new importance to them in the second half of the sixteenth century, and to consolidate their interests in the region they seized control in 1568-1569 of the three coastal towns of Onor, Barcelore and Mangalore, and established fortresses and factories at each.

Onor—the modern Honavar—was built beside the Sharavati River where it widened about two miles from its mouth into a small lake. A sandbank at the bar prevented ocean-going vessels from entering the harbor, but a channel provided access for small coastal craft. The Portuguese fortress occupied a natural defensive position on a cliff overlooking the river. Within the fortress walls were the homes of thirty Portuguese *casados* with their flourishing gardens of vegetables, vines and coconuts.[5]

Some thirty-five miles upstream from Onor, near the town of Gersoppa, the Portuguese maintained a weighing-house (*casa do peso*), where pepper was weighed and bought. From here the pepper was brought by river to the fortress at Onor for trans-shipment aboard coastal vessels to Goa. Onor became Portugal's principal port of export for Kanara pepper, and for much of the first half of the seventeenth century the single most important Portuguese supply port for pepper anywhere in Asia.[6] However, it was not otherwise a very flourishing trade centre, and the resident casados had to be maintained at government expense since there was little opportunity for them to make an independent living.

About fifty miles south of Onor and a few miles up the Coondapoor estuary was the city of Barcelore, now Basrur. Here the Portuguese had followed their usual practice of building a fortress some distance below the Hindu town so as to command the river approaches. According to Bocarro, who was government archivist at Goa between 1631 and 1643, Portuguese Barcelore was a thriving trade centre exporting rice, textiles, saltpeter and iron from the hinterland, and importing corals, exotic piece goods, horses and elephants. A church, and houses for the captain, various civil and ecclesiastical functionaries and thirty salaried Portuguese casados, were located within the fortress. Another thirty-five casados with their families lived in a Portuguese settlement, surrounded by mud walls, a musket's shot from the fort.[7]

The most southerly of the Portuguese strongpoints on the Kanara coast was the fortress at Mangalore, a town situated near the mouth of the Netravali River, and between the territories of Banguel (Bangher) to the north and Olala (Ullal) to the south, independent petty princedoms until the rise of Ikkeri. Mangalore fortress was built on the site of an old Hindu temple, was a solid square construction with a bastion at each corner, and contained the usual church, captain's accommodation and storehouses. The nearby Portuguese town was protected by a wall, and included stone dwellings for thirty-five casados.[8]

The principal export from both Barcelore and Mangalore was paddy rice, used for provisioning the viceregal capital, its outports, and its fleets. According to Bocarro, Barcelore and Mangalore together comprised "the sole granary from which Goa, Malacca, Muscat, Mozambique and Mombasa are supplied," and while this claim was exaggerated, rice from these outlets was undoubtedly vital to the Portuguese State of India, especially during drought years such as the early 1630's.[9]

Ironically, although the Kanara towns, particularly Onor and Barcelore, were considered unusually healthy for Europeans, in none did the Portuguese communities compare in size or importance with those of less salubrious localities such as Cochin and Goa itself. Nor, apart from the annual river journeys to Gersoppa and other points to buy pepper, did the Portuguese venture much inland from their Kanara settlements. In contrast to the situation in Malabar, there were virtually no Christians in the Kanara hinterland, proselytizing was discouraged by the nayaks of Ikkeri, and missionaries seldom penetrated the region.

The acquisition of possessions in Kanara by the Portuguese had important implications for their official pepper trade, for it widened the choice of sources available to them, making possible greater flexibility in their buying tactics for this key export commodity. Kanara pepper reduced the dependence of Portuguese buyers on Cochin and other Malabar outlets, and allowed the bulk of orders to be shifted from one regional source to the other as political or other exigencies demanded. Until the rise of Ikkeri the Portuguese purchased Kanara pepper either from private dealers or from petty rulers in the pepper country whose goodwill and cooperation were encouraged by gifts, formal treaties and occasional shows of force.[10] Both the king of Portugal and his viceroy at Goa were at pains to maintain personal contact with the rajahs, nayaks and ranis concerned, exchanging courtesies with them by letter.[11] It was therefore understandable that when several of these petty rulers were threatened by the expansion of Ikkeri in the early seventeenth century, they should appeal to Goa for assistance.

The initial response of the Goa authorities to Ikkeri's expansion was cautious. They perceived the danger to Portuguese commercial and political interests but were reluctant to risk open confrontation with Venkatapa. They therefore tried at first to limit their intervention to verbal encouragement to the threatened princes, attempts to manipulate them into defensive alliances and diplomatic appeals to Ikkeri.[12] However, these measures were not successful, and by 1618 the Portuguese had become directly involved through the provision of troops, artillery and munitions to their protegé and close neighbor at Mangalore, the king of Banguel. Despite initial successes, they suffered a serious defeat near Mangalore in late 1618, losing their local commanders, Francisco de Miranda Henriques and Luis de Brito de Melo.

Under heavy pressure elsewhere in Asia at this time and lacking the military resources to sustain a prolonged war against Ikkeri, the Portuguese were obliged to seek peace. They forced the king of Banguel to accept unfavorable terms from Venkatapa and his ally, the rani of Olala, and in 1623 they dispatched a peace embassy to Ikkeri itself.[13] Pietro della Valle, who took the opportunity to accompany this mission, was highly critical of its attitude, which he thought excessively conciliatory. "Behold by whom are routed in India the armies of the King of Spain," he wrote, in contemptuous reference to the rani of Olala, whom he met while walking down the single street of her town, adding that she seemed "rather a dirty kitchen-wench, or laundress, than a delicate and noble queen."[14] His comments underline the inability of the Portuguese to intervene effectively in the internal politics of the Kanara region by the early seventeenth century.

Following his conquest of the Kanara pepper lands Venkatapa proceeded to insist that the Portuguese buy most of their pepper from him at what seemed to them excessively high contract prices, a policy later continued by Vira Bhadra. Arranging these contracts as advantageously as possible, and ensuring that their terms were fulfilled by the rulers of Ikkeri, whom the Goa authorities considered particularly fickle, became the major concern of the Portuguese in their relations with this princedom during the later Hapsburg years (c. 1620-1640).[15] Therefore, although Kanara pepper was still acquired by the Portuguese in this period, it was bought at less competitive prices, and in an atmosphere of greater uncertainty.

The conquests made by Venkatapa and held by Vira Bhadra meant that the three Portuguese possessions in Kanara had become enclaves in Ikkeri territory. In 1629, to strengthen the most exposed of these possessions, the count of Linhares, who had recently arrived as the new viceroy, established a second fortress at Cambolin located on a tongue of land near Barcelore, but closer to the bar. However, because Vira Bhadra resented this new fortress, it proved an obstacle to harmonious relations with Ikkeri. Cambolin was supposedly of superior defense capability to Barcelore, possessed a more accessible water supply, and was more favorably located for controlling both the adjacent paddy lands and the sea approaches. It was apparently Linhares' intention to transfer the Barcelore factory to this new site, but in practice both places were occupied simultaneously and continued so until abandoned to Ikkeri forces in 1653-1654, despite prolonged discussions on their respective merits by the viceroy's council at Goa.[16]

South of Kanara, from Mount Deli to Cape Comorin, was the region known to Europeans as Malabar. This was the traditional heartland of the Indian pepper country, and during most of the sixteenth century, and more intermittently in the seventeenth, was the principal source for Portuguese pepper purchases in India. Marriages and alliances between various ruling families of the borderlands of Kanara and Malabar linked the two regions politically.[17] However, unlike Kanara, Malabar during the Hapsburg period was not seriously threatened by landward invasion or by the expansion of any local power to upset the existing political balance. The Malabar region was, and long had been, divided into a series of petty kingdoms and lordships, none of which exercised authority over more than a few hundred square miles of territory. It possessed no power comparable to the major Moslem states of the Deccan or even the Hindu princedom of Ikkeri.

Among the Malabar kingdoms Cannanore, Calicut, Cochin, Quilon and Travancore were the largest and most powerful in the Hapsburg period, but at least twenty-five other political entities of some importance existed, most of them subject to a greater or lesser extent to one of the five larger units. Most significant of the lesser lordships were Cranganore and Porca (Purakkad) which were loosely subject to Cochin, Cale Coulam (Kayam Kulan) subordinate to Travancore, and Ponnani, a town on the river of the same name, and a satellite of Calicut.[18]

The territories of Cannanore, a green and fertile region which formed the most northerly of the leading Malabar states, stretched from the Kanara border near Mount Deli inland to the foothills of the Western Ghats, and southward to the river Kottakal a few miles down the coast, where they adjoined the lands of Calicut.[19] During the sixteenth and early seventeenth centuries Cannanore was a flourishing trading port and as late as the 1670's, though then in decline, could still be described as populous and "inhabited by rich Mahometan merchants."[20] These merchants were the Indianized descendants of Arab immigrants of the eighth century onwards, and were known locally as Moplahs. In Cannanore, as in all the Malabar towns, the Moplahs dominated trade. Since about 1525 the Ali rajah, hereditary leader of the Cannanore Moplahs, had exercised administrative authority within the town, although the de jure ruler was a Hindu prince, the *kollatiri*. The Portuguese conducted most of their formal business in Cannanore with the Ali rajah and his officials.

Europeans who visited Cannanore during its prosperity praised the richness of its famed bazaar, noted for textiles, pepper, ginger and cardamom.[21] Cannanore pepper was reputedly the best in Malabar and in the sixteenth century Portuguese factors sometimes bought supplies in the town. However, between 1587 and 1605 official purchases were made only in 1593, probably because the Portuguese were unwilling to increase the Ali rajah's by then traditional pay-off per quintal of pepper exported. In 1606 when supplies from elsewhere in southwest India and from Malacca were becoming scarce, a compromise price agreement was made, and official Portuguese buying resumed. Thereafter purchases remained at best irregular, and in the Hapsburg era Cannanore was never a consistent major supplier of pepper for the official export trade to Lisbon via Goa.[22] The Portuguese possessed a small fort at Cannanore, their first on the Malabar coast. In the 1630's this fort and the adjoining area contained about forty Portuguese casados and their families, with their churches, houses and vegetable gardens. These settlers formed too insignificant a group to offer any serious challenge to the predominance of the local Moplahs in the commercial and political life of the town.[23]

Adjoining the territories of Cannanore lay those of Calicut (Kozhikode), the largest of the petty kingdoms of Malabar. It was composed of a stretch of coast about a hundred miles long from the river Kottakal in the north to the river Cranganore in the south, with an average breadth of approximately sixty miles. Under its rajah, the Samorin of Calicut, were nine subordinate princes, among them the ruler of the coastal lordship of Grimgal Namboory where major pirate nests were located. The Samorin also claimed suzerainty over the other rulers of the Malabar coast, with the exception of the rajah of Travancore and possibly the kollatiri of Cannanore. However, this claim was strongly disputed by the rajah of Cochin who, backed by the Portuguese, had successfully maintained his independence since the early sixteenth century. Although contemptuously described by a contemporary Englishman as "a naked Negro, but not a little puft up by being the principal Bracman," the Samorin was still the most important ruler in Malabar in the early decades of the seventeeth century.[24]

Like the other coastal cities of Malabar, Calicut and Ponnani were centres for coastal and oceanic trade, and were also located in highly productive agricultural areas. Calicut was one of the most cosmopolitan of the Malabar ports, and Moplahs, Jews and Gujaratis controlled most of its commerce. Just as Calicut's commercial predominance in Malabar had been broken in the sixteenth century by the Portuguese promotion of Cochin, so a resurgence in its prosperity in the early seventeenth century was made possible partly through the decline of Portugal's power in the region. The French traveller and writer François Pyrard, who spent several months in Calicut in 1607-1608, described the place then as "the busiest and most full of all traffic and commerce in the whole of India," and the Dutch *predikant* Baldaeus in 1672 still referred to it as the capital of Malabar. Pepper, cloth and precious stones were probably the most important commodities exported.[25]

During most of the sixteenth century the Portuguese had striven to overawe Calicut, but their occasional raids and punitive expeditions, and their forts at Calicut itself (1522-1525), at nearby Chalyam (1531-1571), and at Ponnani (begun 1535 and never completed), only partially and intermittently achieved this objective. From the late sixteenth century their policy had had to be modified to suit the new political circumstances, and emphasis was increasingly placed on avoiding hostilities. Pyrard, following his stay in Calicut, reported that the Portuguese "by divers presents cultivate as best they can the friendship of this king, whom they fear more than any other," and in fact from 1599 Calicut and the Portuguese were normally at peace. It is clear, however, that Portuguese efforts to persuade the samorin to withhold his tacit support for the Malabar pirates, who lurked in the minor harbors and creeks of this coast and seriously hampered seaborne trade, were not very successful, as della Valle's description of stolen Portuguese goods for sale in Calicut in 1623 bears witness.[26]

During the early seventeenth century the Portuguese maintained a trade factor in Calicut, whose principal function was to issue *cartazes* (licences) to Calicut country vessels sailing north or west through seas where the arm of Portugal, though weakened, was still strong enough for its self-proclaimed right to control maritime trade to be taken seriously. Official purchases of pepper for shipment to Lisbon via the Cape were not normally made at Calicut, but the Portuguese community—centred on the factor, the secretary, and the Jesuit church with its two priests, and swelled by the presence of various Eurasians, Indian Christians, and other hangers-on—traded privately in a variety of products.[27]

South of Calicut lay the rival kingdom of Cochin, stretching along the coast from Cranganore in the north to Porca in the south, and including the lordships of nine subordinate rajahs in its territory. The rajah of Cochin, who claimed descent from the ancient south Indian Perumal dynasty dating back to the ninth century A.D. and resented the pretensions to suzerainty of the more powerful samorin of Calicut, had allied himself to the Portuguese since the time of Cabral's visit in 1501. The alliance, which proved lasting, made Cochin the principal Portuguese factory on the coast of Malabar. It also enabled the Cochin rajahs to assert their independence of Calicut, though not without much intermittent fighting and skirmishing over the succeeding decades.

The old town of Cochin, called Upper Cochin (*Cochim de cima*) by the Portuguese, was situated about four miles up an island-studded channel separated from the open sea by a sandspit. It marked an important point of access for smaller shipping to the lagoons and backwaters of this part of Malabar. The Portuguese-administered town was Lower Cochin (*Cochim de baixo*), about a mile and a half downstream.[28] The rajah's city of Upper Cochin had the usual facilities of the Malabar towns, including an impressive bazaar. In early Dutch times it was described as "very populous" and was said to possess broad streets and some fine buildings, including an exceptionally large tank. The rajah's authority prevailed everywhere outside the walls of Portuguese Lower Cochin, even in the nearby countryside where the Portuguese possessed farms and orchards.[29]

The population of Upper Cochin was swelled not only by Moplahs and other Moslems, but also by sizeable communities of Jews and Saint Thomas Christians. The Jews of Cochin were described by both Linschoten and Baldaeus as numerous and wealthy, and the fact that the rajah followed a policy of religious toleration enabled them to go about their business freely, unhampered by the persecuting restrictions typical of contemporary Catholic Europe and its overseas possessions.[30] The Saint Thomas Christians, though increasingly harassed from the mid-sixteenth century by romanizing missionaries and the Goa Inquisition, often provided the Portuguese factory at Cochin with supplies of pepper for export.

Lower Cochin, also known as the city of Santa Cruz, was built on a low-lying stretch of sand near the mouth of the channel. It obviously impressed Viceroy Linhares during his visit there in March 1631 for he described it as large and beautiful with "great rows of houses."[31] In fact Cochin was second only to Goa in size and importance among the Portuguese towns on the west coast of India, and in the 1630's its population supposedly included 500 casados, 300 of them Portuguese and Eurasians, and the remainder Indian converts. Portuguese Cochin boasted a city council (*câmara*), a cathedral with its presiding bishop, a customshouse, five parish churches, and several convents.[32] Outstanding among the latter were the three-storied Jesuit college, and the Jesuit church with its "lofty steeple and a most excellent set of bells."[33] Though the city was partly surrounded by a wall with several bastions, in the early seventeenth century it lacked sufficient artillery for proper defence. It possessed a good deep-water harbor, but the entrance was obstructed by shifting sandbanks, so that larger vessels and particularly Portuguese carracks were forced to anchor off the bar where they were exposed to enemy attacks. Peter Mundy observed after his visit in 1637 that Portuguese Cochin was "a very large towne which with the monasteries maketh a goodly prospectt," adding that the buildings were "very faire and uniforme" but less numerous and less grand than those of Goa.[34]

The forests of Cochin were one of the principal sources of teak, poon and other timbers used by the Portuguese for shipbuilding, and the local shipyards were among the most important available to them on the western coast of India. Cochin was prominent as an entrepôt for merchandise from the east coast of India, Sri Lanka, Mozambique, the Far East and other borderlands of the Indian Ocean. In the sixteenth and early seventeenth centuries it was also the major centre for the purchase of Malabar pepper for export to Lisbon, and the principal port of lading for the carracks that carried this product home. In the last decades before the Dutch and English forced their way into the Indian Ocean trade, up to four or five carracks might call annually at Cochin to load "with pepper and drugs and with all other oriental merchandise and riches" to carry back to Portugal.[35]

However, whereas Cochin was described as a city "of great prosperity and growth" at the beginning of the Hapsburg period in 1582, fifty years later António Bocarro was lamenting its alleged reduction to "extreme poverty through lack of trade."[36] This complaint was perhaps exaggerated, for in 1629 the Portuguese India Company still bought a major proportion of its pepper at Cochin: 3,321 heavy quintals, which comprised about 31 per cent of the 1630 consignment for Lisbon.[37] Moreover, whatever the fate of the official Portuguese spice trade, private trade to and from Cochin, mainly with other ports on the west coast of India and across the Arabian Sea, continued to prosper in the early seventeenth century, and was sufficiently flourishing in the decade after 1610 for one anonymous writer to state that "today there are no rich men in India except in Cochin."[38] Nevertheless, because of the increasing menace of Dutch seapower, Cochin was virtually abandoned as a port of lading for Lisbon-bound carracks in the last years of the sixteenth century, though still used on rare occasions for this purpose until 1611. A decade later the crown raised the possibility of building a protective mole at Cochin as a preliminary to reviving it as a port for large ships, but the idea was rejected as impracticable. From 1611 onwards all Malabar pepper bought by the Portuguese was trans-shipped to Goa in small swift coastal craft, and only on reaching the viceregal capital was it loaded on the ocean-going carracks for Lisbon.[39]

In addition to their establishment at Cochin the Portuguese had maintained a fort at the satellite city of Cranganore (Kodungalloor), close to the border with Calicut, since the first decade of the sixteenth century. Built beside the Cranganore River, this fort was of strategic importance because it commanded the main invasion route from Calicut to Cochin. In late Hapsburg times it was supposedly manned by one hundred soldiers, some of whom were Saint Thomas Christians. In the nearby Portuguese settlement lived about forty white casados.[40] The Cranganore hinterland produced pepper in abundance, and the local Saint Thomas Christians were "very great friends to Portuguese merchants to whom they sell their pepper and other aromatic products."[41] Nevertheless, the Portuguese did not use Cranganore itself as a regular pepper buying centre, largely because of the proximity of Cochin, whose rajah objected to their utilizing a rival mart so close to his own doorstep. Cranganore's main importance to the Portuguese was rather as a missionary centre aimed at converting the Saint Thomas Christians from Nestorianism to Roman Catholicism. This conversion was formally achieved at the Synod of Diamper in 1599. The Roman Catholic bishopric of Angamale with headquarters at Cranganore was then formed, and raised to archiepiscopal status in 1608. In late Hapsburg times both the Franciscans and the Jesuits had colleges in the Cranganore region, the Franciscan college in the town itself and the Jesuit in nearby Vaipicotta.[42]

From the Cochinese border at Porca to the southernmost point of India at Cape Comorin stretched the territories of the remaining ruler of importance in Malabar, the rajah of Travancore. His dominions included the lordships of eight significant subordinate rulers, over most of whom his authority was little more than nominal. Eventually, however, the Travancore dynasty was to emerge as the strongest in Malabar, most of which it absorbed and united under Rajah Martanda Varma in the mid-eighteenth century.[43]

The Portuguese interests in Travancore were centred on the subordinate lordship of Quilon, where the leading port of the region was located. This city was described by Baldaeus as "the least among the Malabar kingdoms." As in the case of Cochin, though on a smaller scale, Quilon consisted of a port section, where the Portuguese fortress and other facilities were located, and the Indian town situated a little further inland. It was remarkable for "having no less than seven churches, some very good houses, and many thousands of trees, especially towards the seaside," and was reckoned to be outstandingly healthy.[44]

Quilon enjoyed not only direct access to the sea, but communications with Cochin for small shallow-draft vessels via the inland waterways. Because Portuguese influence was concentrated at Quilon, the Portuguese often referred to it as though it were the leading power of south Malabar, but António Bocarro rightly distinguished between the petty local ruler and the rajah of Travancore, "who is a much greater king."[45] The Portuguese and Eurasian population of Quilon was small, and in the 1630's was supposed to have numbered about sixty, including young bachelors as well as casados.

Because Quilon was an important centre for the purchase of pepper for export to Lisbon, it had its own Portuguese trade officials. Nevertheless, since the mid-sixteenth century general supervision of Portuguese commercial activities at Quilon had been exercised by the treasurer (*vedor da fazenda*) at Cochin.[46] Occasional carracks had been loaded with pepper at Quilon in the later sixteenth century, and further consignments were sent to Cochin or Goa for trans-shipment to Europe. When differences with the nayaks of Ikkeri hampered purchases in Kanara in the 1620's Portuguese pepper exports from Quilon for a time exceeded those from all other ports, despite the diverting of much of the crop eastwards across Travancore to buyers in Coromandel and beyond. In 1629, 5,686 heavy quintals of Quilon pepper were purchased by the Portuguese India Company, constituting 52 per cent of exports to Lisbon in 1630, and thus making the south Malaba port easily the largest single supplier in the first season of the company's existence.[47]

Finally, in addition to their settlements at Cochin and Quilon, the Portuguese maintained direct and indirect contacts with many of the petty lords and princelings of the neighboring country. Among these were the so-called pepper king (*rei da pimenta*), who ruled the coastal lordship of Vettakenkur a few miles south of Cochin city, and the rulers of Porca, Mangatti and Parur, also in the Cochin region. Near Quilon were the rajah of Cale Coulam, the rajah of Marta, and several others, while in central south India ruled the more important and powerful nayak of Madura. These rulers were known collectively to the Portuguese as the neighboring kings (*reis vizinhos*).[48] Since most of the orchards for Malabar pepper were located in or accessible through their territories, it was desirable for the Portuguese to keep these rulers well disposed, which they sought to do through a combination of coercion, flattery and presents.

2 / Goa and Portuguese Trade

During the Hapsburg period Goa was the local capital for an empire which embraced about fifty official settlements and possessions ranging in importance from major entrepôts like Malacca to small outposts such as Sena on the Zambezi, where by the 1630's there was neither a fortress nor even usable cannon. In most of these places Portuguese political control did not extend beyond the immediate vicinity of the base, although adjoining a few, including Damão and Bassein on the coast of Gujarat, Colombo in western Sri Lanka and Goa itself, were significant tracts of subject territory.[1] Under normal circumstances only Goa itself traded and communicated directly with Portugal, all other possessions being obliged to do so indirectly through the viceregal capital.

The city of Goa was built on the island of Tissuary located approximately midway down the west coast of India. It was an almost ideal position for trade purposes, with the textile and indigo-producing areas of Gujarat to its north and the pepper lands of Kanara and Malabar to the south. Roughly in the shape of a triangle with its apex pointing towards the sea, Tissuary was a tongue of land of some fifty square miles formed between the Mandovi River to the north and the shallower Zuari River to the south. To the east a narrow creek flowing from the Mandovi to the Zuari linked the two and completed Tissuary's isolation as an island, but this channel was so slight in places that in the dry monsoon season a man could ford it easily, the water only reaching his knees.[2]

In addition to Tissuary Island, the Portuguese also possessed the neighboring mainland districts of Bardez on the north bank of the Mandovi and Salcete on the south bank of the Zuari, together with the intervening islands, of which Juá, Chorão and Divar were the most important.[3] Bardez was almost enclosed by the sea to the west and by the rivers Mandovi, Chapora and Mapusa on the south, north and northeast. It could be conveniently entered only from neighboring Bijapur through a narrow gap called the Bardez Gates. Salcete also was partially protected by bordering rivers, although here the pass into Bijapur was considerably wider and less easily defensible. On the eastern side of Tissuary Island the shallow Mandovi-Zuari channel marked the frontier with the Shahi sultanate, and this border had been reinforced by a wall. The total Portuguese dominion in the Goa region in the early seventeenth century amounted to approximately 275 square miles.

Much of the island of Tissuary is composed of lateritic platforms of poor fertility, but these are interspersed with alluvial flats where in the seventeenth century—as today—coconuts, rice and millet flourished. Seventeenth-century visitors to Goa were often impressed by the abundance and excellence of the household gardens and orchards in and near the city, and commented on the perpetually green appearance of the nearby countryside. On the other hand, there was little livestock apart from a few sheep and goats, and the island had neither the agricultural nor pastoral resources to feed the city's population, which was therefore dependent on imported food.[4] Paddy, coconuts and other crops from the outlying regions of Bardez and Salcete supplied some of this need, but most of Goa's food was brought in from neighboring Bijapur, from the Kanara ports to the south, or from Damão, Bassein and other supply centres to the north.[5] This dependence caused much concern to the viceregal government when Portuguese seapower was challenged by the Dutch in the first decades of the seventeenth century, and local Indian rulers grew proportionately less willing to accommodate Portuguese needs. In the early 1630's the problem reached crisis proportions as widespread famine made northern supplies temporarily unavailable, and at the same time political differences with Ikkeri obstructed the purchase of foodstuffs from alternative sources to the south.[6]

The problem of supplying Goa with adequate food was compounded by the size of its population, said to have been remarkably large for its time. It has been claimed that Goa had about 225,000 inhabitants around 1600, and thus ranked alongside London and Antwerp as one of the largest metropolises of the age.[7] However, in the absence of clear evidence such a high figure must be viewed with some scepticism. The Frenchman Jean Mocquet, who spent nine months in Goa between May 1609 and January 1610 and is one of the few contemporaries to comment on the size of the city, wrote that Goa "may be about as big as Tours."[8] If his suggestion is correct, and Goa was indeed about the size of a large French provincial town, then its population must have been a good deal smaller in Mocquet's day than is usually assumed. Bocarro estimated the native population of the Goa territories in the 1630's to be approximately 100,000, although it is unclear whether this included the inhabitants of the city as well as those outside its confines. About 60 per cent of native Goans—or *Canarins* as the Portuguese called them—were by this period Roman Catholics, and the remainder Hindus of varying status. Most of this Goan population, both Catholic and Hindu, consisted of farmers and fishermen who lived with their families in villages by the rice paddies and coconut groves, or on the coast and river banks.[9]

Of the many non-Goans living in Goa, slaves apparently comprised the largest single category. Indeed, it was alleged in the early seventeenth century that most of the Goa population consisted of slaves, and their number was described as infinite.[10] The typical Portuguese casado could be expected to own about ten slaves, while Goan householders in the city were also normally slaveholders. Most of the slaves were seemingly Negroes from Mozambique and other parts of East Africa, but some were Asiatics. They were variously used as guards, concubines and personal servants, or were put out to work at trades and handicrafts, for the profit of their masters. That slaves were also exported or re-exported to Portugal is fully corroborated in the manifests of some homeward-bound carracks from Goa to Lisbon.[11]

Immigrants from the neighboring states of India, especially Bijapuris and non-Goan Kanarese, and Gujaratis and other Hindus from the northern portion of the west coast, formed a significant minority in Portuguese Goa. Some were craftsmen—goldsmiths, coppersmiths, carpenters and barbers—but many were traders, dealing especially in cloth and foodstuffs. Bengalis, Arabs, Persians, Armenians and Jews were also well represented in the Goa community—Malays, Chinese and Japanese rather more rarely. Needless to say, no precise figures can be given for the numbers of these foreigners, many of whom were probably only temporary residents.[12]

Sixteenth and seventeenth-century sources are more informative about the European and culturally Portuguese Eurasian populations of Goa. Apart from the relatively small numbers of *fidalgos*, lawyers and bureaucrats who monopolized most of the more lucrative political and military offices in the *Estado da Índia*, these people may be divided into the three categories of casados, *soldados* and religious. The vast majority in each category was naturally composed of Portuguese or Eurasians of part Portuguese descent, although Italians, Castilians,

Germans, English, Dutch and French were present from time to time in small numbers. Second or third generation Portuguese born in Goa usually had some Asian blood in their veins since the casados were commonly married to Eurasian women, but the annual arrival of young bachelors on the carracks from Lisbon ensured that there was always a proportion of full-blooded European Portuguese in the community.

The casados, who formed the backbone of the Portuguese population in Goa, were mainly former soldiers who had married and settled locally, or the descendants of others who had done so in the past. Although some Portuguese women accompanied their husbands to India, and female orphans were occasionally despatched there to find husbands, the numbers in both categories were probably very small.[13] In practice, taking a Eurasian or Indian wife was normal for most Portuguese settlers in India.

The casados of Goa appear to have represented a fair cross-section of Portuguese society. They included government officeholders as well as landholders, merchants and lesser officials, or people who combined two or all three of these occupations, and a core of artisans and craftsmen who, like their social superiors, were represented in the city *câmara*.[14] The number of casados in the city probably remained fairly stable from about the middle of the sixteenth century into the first or second decade of the seventeenth, averaging nearly 2,000. However, by the 1630's their numbers had apparently dwindled to about 800, a figure which reflected the marked decline of the city after the siege by the Bijapuris in 1570 and the subsequent arrival in the region of the Dutch and English. There were also a few married Portuguese resident in the Goa territories outside the city—especially in Bardez, where most lived on *fazendas* held from the crown in return for a quitrent. The Portuguese casado population of the outlying settlements in Asia was probably smaller than that of the Goa territories, although Viceroy Linhares' assertion in 1634 that there were not even 1,000 such settlers in the whole *Estado da Índia* seems hardly credible.[15]

The term *soldado* was applied not only to European soldiers shipped out to India from Lisbon, but to all unmarried Portuguese men in Asia eligible and able to bear arms, and also to similarly qualified Lusitanized Eurasians. The total number of soldados in Goa at any particular time must have varied greatly according to circumstances, being highest when the carracks arrived from Europe and lowest when the various coastal fleets and military expeditions were in operation during the months of the dry monsoon. Pyrard declared that he had seen 4,000-5,000 Portuguese and Eurasian soldiers in Goa. However, Mocquet reported that there were about 1,500-2,000 in the city "according as the fleets arrive," adding that he personally had seen "a Muster-General of all the inhabitants bearing arms, as well the Portugals as the natives and Indians, and [they] were found to be about 4,000." Two decades later Bocarro wrote that the number of soldiers in Goa fluctuated widely, but was "sometimes about 1000." This figure, which has some corroboration elsewhere, may be taken as an approximate indication of the operational manpower available to the viceroy at the start of an average year, in the last decade or two of Hapsburg rule.[16]

The authorities at Goa could also call upon the local Goan Christians for military service, but they had a mixed reputation, some Portuguese regarding them as "not people who face up to European enemies, or to the Turks."[17] However, sepoy troops were on occasion utilized, though usually for auxiliary purposes such as scouting and trail-cutting, and on particular undertakings for limited periods.[18] In 1631, for example, Viceroy Linhares strengthened his expeditionary forces by imposing manpower levies of 400 and 500 men respectively on the Goan territories of Bardez and Salcete. When certain wealthy individuals in Salcete complained that they had been unable to acquire the numbers of native Canarins needed to meet their quotas, at the going price for a substitute of 100 *xerafins* per head, the viceroy told them they should obtain Negroes instead who were in any case cheaper.[19] Negroes, whether slave or free, were in fact considered more aggressive and a much more reliable military standby in times of emergency than Canarins, though recruits from both peoples were used as available.

In practice there were never enough soldiers of any kind in Goa during the Hapsburg years to meet the needs of the state, as the viceroys and governors perennially complained.[20] Among the principal reasons for this were the inability of the Lisbon authorities to despatch the required number of men from Portugal herself, the drawing up of fraudulent registers of recruits at the *casa da Índia,* the enlistment of underage boys instead of mature men, and the high mortality rates on the voyages out. Most of these abuses were European in origin, but conditions and attitudes in Goa helped to compound their effects and to ensure that only a relatively small proportion of Portuguese soldiers who did reach India actually served there for very long. The death rate among European troops was extremely high in Portuguese India, where tropical diseases abounded. More than 500 soldiers who had reached Goa aboard Nuno Álvares Botelho's galleons in 1624 in a healthy condition died subsequently in the city hospital, allegedly because the patients there were "badly cured."[21] By the end of the century, if the Augustinian Agostinho de Santa Maria is to be credited, 25,000 soldiers had died within the walls of the Royal Hospital alone.[22] In addition, the Hapsburg years saw not only disease but also the frequent and desperate fighting with the Dutch and other enemies, exacting a heavy toll.

The number of soldiers available to the Portuguese in India was further reduced by a high rate of desertion. Conditions of service in Goa were highly unattractive, especially during the wet monsoon season when inactive soldiers, unhoused and often unpaid, reportedly went about "stripped and naked on the roads, seeking alms."[23] Moreover, since many recruits were actually "scoundrels from the prisons of Portugal" they felt little compunction in transferring their services to an Asian ruler who gave hope of a better life. In fact, so many Portuguese soldiers absconded in this way that perhaps not more than one-fifth of those who embarked at Lisbon actually served at Goa, while in 1627 an estimated 5,000 Portuguese renegades were in the employ of local Asian potentates between Bengal and Macassar, "with little hope of remedy for they are accustomed to the free life."[24]

It was also common for men arriving in India from Portugal to evade military service by joining a religious order. In fact, the soldiery provided the orders in India with one of their principal sources of recruits, and the consequent loss of military manpower was at times so serious that the orders were prohibited from accepting any candidates at all from among the soldiers.[25] The count of Linhares, while viceroy at Goa in the 1630's, asserted that religious orders snatched away at least half the soldiers sent from Portugal every year and, with perhaps understandable exaggeration, complained that there were twice as many clerics in India as there were Portuguese laymen. Linhares recognized that the church establishment in Portuguese Asia was a crippling financial burden on the viceroyalty's treasury. In particular, he claimed that the state of India was struggling to support too many grandiose convents full of "lazy religious," even places like Chaul, Damão and Malacca supporting four monasteries each "whereas for confessions and preaching one convent and one religious appear to be sufficient for each of these fortresses." Despite Linhares' critical views and his repeated attempts to curb the clerical recruiters, their activities continued, and by the end of his term of office even this energetic viceroy had given up in despair—"the evil increases . . . the danger is near and I can do no more than weep and cry to Your Majesty." Other viceroys, both before and after, experienced similar frustrations.[26]

The presence of church and clergy must have been extraordinarily visible in Hapsburg Goa. Pyrard described the multiplicity of churches in the city as a marvel, adding that no square, street, or crossroad was without one.[27] Appropriately, the largest building in the city was the great 250-foot-long cathedral of Santa Caterina, commenced in 1562 and still undergoing final construction and embellishment in the 1630's. As metropolitan church for an archdiocese stretching the entire length of the Indies from the Cape of Good Hope to China it was staffed on a grand scale. There were also seven parish churches in the city of Goa and a further sixty-two elsewhere in the Goa territories, each with at least a vicar and parish officials, and some supporting up to four or five beneficed clergy.[28]

Some of the Goa parish churches were fine edifices, but few could equal in scale or magnificence the numerous monasteries, convents and other religious houses clustered in the city, which included some of the largest ecclesiastical institutions in the Lusitanian world. Among the most important was the São Paulo College, the largest Jesuit school in Asia, with a roll of seventy religious, allegedly two thousand students, and extensive endowments. The Jesuit houses of São Roque and Bom Jesus—the latter containing the tomb of St. Francis Xavier, canonized in 1622—housed sixty to sixty-five members of the order between them. The vast Augustinian monastery and the principal Franciscan convent, described as "the handsomest and richest in the world," boasted seventy to one hundred religious each, the Dominican convent sixty.[29] The only regular house for women, the Augustinian nunnery of Santa Monica, which had been founded in 1606, had a statutory limit of a hundred nuns but probably sometimes exceeded this number.[30] Altogether the city and island of Goa were graced with some twenty regular houses, and there seems little doubt that the number of people in orders or in some form of conventual life in Goa easily reached four figures.

Unlike the planned, rectangular cities of Spain and Spanish

America, early seventeenth-century Goa was a complex maze of disorderly streets, squares and lanes, medieval in its haphazardness. In this it was comparable to other Portuguese towns, such as Salvador and Olinda in Brazil, but especially to Lisbon itself, which the Portuguese Indian capital somewhat resembled. Visitors to Goa in the late sixteenth and early seventeenth centuries were often quite impressed by the quality of the architecture. Pyrard was astonished that the Portuguese had managed to construct "so many superb buildings, churches, monasteries, palaces, forts, and other edifices built in the European style" in the century since the island's capture by Albuquerque in 1511. However della Valle, though he considered the buildings in general to be large and good, also thought them somewhat plain and lacking in "ornament or exquisiteness of Art"—a plausible criticism for an Italian passing judgment on the relative austerity of the post-Manueline style.[31]

As in Lisbon, commercial life in the Portuguese Indian capital was concentrated on the waterfront, and on a single large commercial artery which ran for nearly a mile inland at about right angles to the river. This was the Rua Direita, lined with the premises of numerous businessmen, traders and craftsmen, and including facilities for the sale of horses and slaves.[32]

As the capital of Portugal's state of India, Goa was the sole port of arrival for ships engaged in the Portuguese Europe-Asia trade and, after Cochin finally ceased to be used as a subordinate port of lading for the carracks in 1611, the only departure point. At the same time, it was the hub of a large regional trade to other ports in Asia, and to East Africa. To service its extensive shipping activity Goa maintained the only full-scale crown dockyard in the state of India. The *Ribeira Grande* was roughly equivalent to the Lisbon *Armazém* except that it contained not only ship-building and repair facilities, the main naval stores of Portuguese Asia and the arsenal, but also the royal gun-foundry and the mint.[33] The Ribeira Grande was for its day an elaborate enterprise employing many officials, artisans and workmen, and its efficient operation was obviously vital for the successful running of the carreira da Índia.

The trade between Goa and Lisbon, conducted by small annual fleets of unusually large carracks or galleons, called *naus da carreira da Índia,* was regarded by the crown as Portugal's principal interest in the State of India. Although it was mainly a one-way trade, in which the Portuguese paid for their oriental purchases with the proceeds of silver and gold shipped out from Europe, there were also some commercial products exported to Goa, including coral, woollens, linen, various foodstuffs, wines and weaponry. In addition to pepper, the Asian products shipped home from Goa in the Hapsburg period included a wide variety of fabrics from Gujarat, Bengal and other parts of India, cinnamon from Sri Lanka, Chinese silks, Indian indigo and furniture, Chinese and Japanese gilt caskets, and from various parts of South Asia diamonds, cowries, coconuts and rice.[34] Certain key products such as pepper, indigo, and ebony were monopolized by the crown—or, from 1629 to 1633, by the Portuguese India Company—but all other commodities were despatched by private traders, whose consignments comprised a substantial portion of the cargo of each carrack. The total annual tonnage of goods customarily exported from Goa to Lisbon cannot be determined with accuracy. However, the figure for most of the Hapsburg period could hardly have been more than 2,000 tons, since between 1601 and 1640 an average of only two to three carracks or galleons set out on the return voyage each year.[35]

The regional trade to and from Goa was normally very active throughout the year, except from June to September when the wet monsoon made navigation impracticable. Bocarro estimated that the annual total investment in Goa's maritime trade was of the order of 2,850,000 xerafins.[36] This amounts to about fifteen times the value of Portuguese India Company merchandise exported to Lisbon in 1630—a fairly good season. It is a striking indication of the much greater value of Goa's regional trade than its trade with Portugal, though both were in fact in decline during the early decades of the seventeenth century.[37] Most of the regional trade was with other ports of western India, and to facilitate the collection of shipping licenses and to provide security against the local pirates who terrorized the coasts from Gujarat to Malabar, the Portuguese organized several annual convoys of merchantmen escorted by small warships fitted out from the Ribeira Grande.

The main northern fleet, protected by a score or so of armed foists, normally left Goa soon after the onset of the dry monsoon in September or early October, primarily to bring back textiles from the ports of Gujarat. The Portuguese settlements at Diu, Damão, Bassein and Chaul were among the usual ports of call; Cambay, chief port of Moghul Gujarat, was the final destination. As late as the mid-1620's there had also been a second sailing in March, and for some years thereafter convoys of up to three hundred vessels made the voyage twice annually. However, by 1635, as a consequence of Dutch and English competition at Surat, these convoys had been reduced to a single sailing of fewer than forty ships. The Kanara fleets were intended primarily to secure provisions, timber for the Goa shipyards, and pepper from Onor. In 1626 there had been sailings in March, April and October, but a decade later only one. The Cape Comorin convoys left twice yearly. The first sailing was intended primarily to fetch pepper from Cochin and the other Malabar ports, and the second to meet ships coming from Malacca, Bengal and various other points beyond Cape Comorin, and escort them to Goa.[38] Both convoys were reduced in size by the later Hapsburg years as the Dutch proceeded increasingly effectively to hinder Portuguese trade and communications with Asia east of Comorin.

Goa was also the starting point for a number of regular commercial voyages to destinations in Asia and East Africa beyond the west coast of India. These long-distance ventures were usually carried out with a single large ship or a squadron of smaller craft for greater safety, and typically dealt in spices and other luxury products small in bulk but large in value. They were run either directly for the crown on a monopoly basis or, more often, let out under license to private traders. Among the most important trading ventures of this kind remaining by the early seventeenth century was the annual voyage to Sri Lanka, undertaken by small squadrons of galliots or pinnaces, freighted by the crown authorities at Goa. The main purpose of the voyage was to collect cinnamon, although a number of other Sri Lankan products, such as elephants, weaponry, ivory figurines, crystal, mats and straw hats, were also brought back to Goa. In 1637 the Englishman Peter Mundy saw this Portuguese "Cinnamon Fleete" anchored off Bhatkal, in company with the convoy of Kanara and its attending escort.[39]

Also significant in the Hapsburg period were the voyages between Goa and Mozambique, and seaborne trade between Goa and Ormuz till 1621, and after that between Goa and Muscat. For the Mozambique voyage a few pinnaces usually left the viceregal capital at the start of the dry monsoon with cloth, munitions and provisions which they exchanged in East Africa for ivory, ebony and Negro slaves, but especially for the gold which originated in the kingdom of the Monomatapa up the Zambezi River, in modern Rhodesia.[40]

Ormuz was chiefly important as a nerve centre for transit trade to and from Persia, Arabia and the land route to the Eastern Mediterranean on the one hand, and India and southeast Asia on the other. Persian silks and carpets, Arab horses, European silver and manufactures, Indian textiles and pepper, Indonesian spices all passed through this thriving city in its hot and barren wasteland, ships from Goa apparently taking the greatest share. The port of Muscat, of much increased importance after the fall of Ormuz in 1621, gave continued Portuguese access to supplies of Arab horses, to the pearls of Bahrein, and to the overland trade route via Basra to Aleppo, until the surrender of Muscat itself to the Omani Arabs in 1650.[41]

In 1580 the route between Goa and Malacca, and through Malacca to Macao and the rich China and Japan trades, and to Macassar and the East Indies spice trade, was still of major importance in intra-Asian commerce. Malacca itself could then reasonably be described as "the most important and profitable trading centre the Kings of Portugal possess in the State of India."[42] However, in the course of the Hapsburg years the trade and population of this ancient entrepôt declined drastically, crippled by repeated Dutch blockades from 1601 onwards, major attacks from the sultanates of Achin and Johore, and the maladministration of Portuguese captains. By the end of the second decade of the seventeenth century communications between Goa and Malacca had become at best precarious, with Dutch patrols in the Singapore Straits either preventing sailings altogether, as in 1627, or taking heavy toll of vessels that tried to run their gauntlet, as in 1633 when the Goa-Macao traders allegedly lost nine galliots and over a million and a half xerafins.[43]

Nevertheless, the Japan voyage remained highly lucrative until the late 1630's. Normally undertaken before 1618 by a single large carrack that sailed from Goa to Macao via Malacca, and thence on to Nagasaki, it was carried out thereafter by groups of swift pinnaces or galliots, for greater security. As late as 1635 the Japan voyage brought the crown a net profit of 172,000 xerafins, and the trade finally ceased after 1638 only because the shogunate forbade completely all communications with Macao.[44] One explanation for this prolonged prosperity is that the Dutch found it more difficult to harry the Macao-Nagasaki leg of the Japan voyage than the Macao-Goa section. This was probably partly because the Formosa Strait was a more difficult passage for the Dutch to seal than the Straits of Singapore or the

Straits of Malacca, partly because the Dutch attacks on Macao in 1622 and 1629 were decisively repulsed. Moreover, the Dutch could not afford to offend the many influential Japanese, including the shogun, who invested in the cargoes carried by Portuguese Japan ships.

Most local Portuguese, including government and ecclesiastical functionaries, participated in the regional trades, either as active merchants or as investors of money and goods. Some were substantial entrepreneurs like Manuel de Morais Supico, a local director of the Portuguese India Company of 1628-1633, and the resourceful Francisco Vieira de Figueiredo. The latter's trading activities in the 1650's and 1660's stretched over many parts of Asia, embracing Goa, Macao, Macassar, Timor and Batavia—and his commercial associates included the sultan of Gowa, the nawab of Golconda, the Jesuits, English and Moslem traders at Madras and in Indonesia, and even officials of the Dutch East India Company.[45] Frequently, Portuguese trading in Asian seas was carried out in association with Asians, and Gujarati capital in particular was often used to finance Portuguese commercial ventures. Of course, many Asians conducted business on their own accounts at Goa and other Portuguese ports, although the extent of their share in, and even control over, trade in these centres remains uncertain. Most coastal vessels engaged in the predominantly textile trade between Cambay and Goa were apparently Indian-owned.[46]

The Portuguese in Asia not only operated a substantial seaborne trade, but from the early sixteenth century claimed lordship over all Asian seas, required all vessels sailing them to carry Portuguese licenses, and attempted to impose transit dues. In practice this draconian policy could only be implemented in limited coastal areas where Portuguese power was concentrated, and was unenforceable elsewhere. It was most successful with shipping off the west coast of India where the purchase of a license at the nearest Portuguese fortress was standard practice throughout the late-sixteenth and into the seventeenth century. A large proportion of Goa's transit trade, and of its customs revenue, resulted from such coercion.

The existence of this "redistributive enterprise," as it has been called, opened the way for many abuses.[47] Portuguese fortress commanders and officials preyed on local commerce and shipping—in the name of the crown, but actually more for their own individual enrichment. At the same time they assumed, often against crown instructions, exemptions and privileges for their personal business ventures.

Notorious for such behavior was the Portuguese administration at Ormuz, which was matched in this respect by the fortress commanders at Malacca who had an especially unsavory reputation for forcing private merchants to sell goods to them at discount prices.[48]

Overland trade routes from Goa to Bijapur and the Indian interior beyond were also regularly utilized in the Hapsburg period. Following the damaging but ultimately unsuccessful siege of Goa by the armies of the Shahi sultanate in 1570, the sultan had signed a treaty with the viceroy which set the pattern for Portuguese-Bijapuri relations for most of the Hapsburg period. The terms of this treaty (1576) enabled the Portuguese to purchase strategic materials such as shipbuilding timber, saltpeter and horse fodder from the neighboring mainland, as well as essential supplies for the Goa population including meat, wheat, rice, vegetables, fruit and charcoal.[49] Goa merchants traded in such products on a year-round basis with and through Bijapur, also importing textiles and various luxury goods such as diamonds from the mines of Golconda, Indian spices and bezoar stones. In exchange they provided Arab horses, drugs and spices from elsewhere in Asia, and gold, silver, coral and manufactured goods originally imported from Lisbon or via the overland routes from other parts of Europe.[50]

Landborne commercial traffic from Goa through Bijapur was seldom disrupted during these years, although occasional strains occurred in relations between the two powers, especially with the growing presence of the Dutch and English on the west coast of India in the early seventeenth century. Portugal's difficulties with its European competitors were understood by the Bijapur sultans, who probably hoped for support from the Protestant companies to recover their lost Goa territories. In 1628 the Bijapur governor of Karrepatan invited the English to settle in his town, promising to supply them with 2,000 tons of pepper per year. At about the same time the Dutch tried to establish a factory at Rajapur, and in 1638 they moved into Vengurla, which they used as a supply base for their periodic blockades of Goa.[51] Fears of a Bijapuri attack on Bardez were current in the viceregal capital in the late 1620's and early 1630's, but skillful Portuguese diplomacy and a certain amount of luck ensured that such a potentially disastrous development was avoided.[52] In reality, Bijapur was probably not capable of launching a serious invasion of Goa in late Hapsburg times, being too preoccupied with the Moghul offensive against its northern and eastern borders, which culminated in the siege of Bijapur city itself by Shah Jahan's forces in 1631.

The trade network that emanated from Goa during the Hapsburg period, despite the closure or decline of some routes, proved remarkably resilient. At the same time contemporary observers—including detached foreigners like Pyrard, as well as involved Portuguese like Bocarro—realized that while the importance of Goa as an entrepôt might be prolonged by one expedient or another, the trend was at best towards a more complex system of rival trade centres with the Portuguese enjoying but a share, and a diminishing share, of their previous markets. They were living in the twilight of an empire whose fortunes were steadily dimming.

3 / The Structure of the Pepper Trade

Sources documenting the structure of the Portuguese pepper trade in India in the sixteenth and seventeenth centuries are disappointingly meagre. It seems that few Europeans actually visited the pepper orchards of Malabar and Kanara in these periods, and consequently such descriptions as were written of the cultivation of pepper and of the process by which it reached the markets of the coast were mostly brief, superficial and second-hand. The procedure after the pepper came into the hands of the Portuguese factors in Malabar and Kanara is somewhat better documented, and fairly detailed statements of account are available at least for a few isolated years of the Hapsburg era.[1]

One of the earliest European writers to discuss the pepper trade was the sixteenth-century geographer Duarte Barbosa. He included in his survey of the western coastlands of India a short description of the middlemen who in the early sixteenth century purchased pepper in the growing areas, and brought it down to the Malabar ports for resale to the exporters. Garcia d'Orta also discussed pepper in his treatise on the drugs and simples of India, published at Goa in 1563, but was more interested in its botanical and pharmaceutical properties than in its cultivation or trade.[2] Among non-Portuguese Europeans who visited India neither Linschoten nor Pyrard, both of whose accounts of the pepper orchards were superficial, provided much enlightenment. More informative was the Englishman Peter Mundy, who actually passed through some of the pepper country during his journey to Ikkeri in 1637, and was able to write from first-hand observation. However, none of these sources can be regarded as more than supplementary, and, were it necessary to rely on them alone, even tentative reconstruction of the Portuguese pepper trade in India would not be possible.[3]

Fortunately, there also exists a detailed treatise on the trade by Francisco da Costa, an obscure Portuguese trade official at Cochin in the late sixteenth and early seventeenth centuries.[4] What little is known about da Costa is derived either from this treatise or from the supplement added by his brother, Luís. Francisco had become scribe (*escrivão*) of the Portuguese factory at Cochin in about 1582 and served in this relatively humble post for the next thirty years, until his death in 1612. The treatise itself was completed in 1607, and was based on a quarter-century's experience in the trade. It is a largely technical work consisting principally of a history and survey of current procedures, price fluctuations, weight standards, losses in weight (*quebras*) to which pepper was liable after purchase, and other relevant data for each of the Portuguese factories in Asia. Also included was a short discourse on the cultivation and harvesting of pepper, based on information from merchants and farmers.[5] It described the methods of agriculture used in the orchards and provided further details on the purchase and weighing procedures, and on the various deceptions for which Portuguese officials had to be alert when transactions were in progress.

After Francisco's death, Luís da Costa, who also seems to have had experience of the India pepper trade, brought his brother's treatise up to date as far as 1622 and was probably responsible for adding a copy of the standing orders for the pepper trade issued by the India Council (*Conselho da Índia*) in 1612.[6] The entire work with its various amendments and additions was forwarded to the crown, but apparently aroused little interest. It remained as an unpublished codex in the archives at Simancas until 1963.

Although the Costa treatise is a valuable addition to the limited information available on the pepper industry of southwest India, it still leaves many questions inadequately answered. For one thing, Francisco da Costa himself seems never to have visited the pepper-producing areas in person, despite his long association with the trade; much of his information was therefore second-hand. He wrote little or nothing about the system of land tenure in the orchards, or the identity of owners and cultivators, and barely touched upon the middlemen, their business procedures, and their profits. On these and related matters a few hints can be gleaned from official correspondence and reports, and from other contemporary materials including surviving records of the Portuguese India Company of 1628-1633. Also, inferences can be made from comparisons with post-Portuguese accounts of the region. However, unless and until more thorough evidence is forthcoming, many aspects of the cultivation of pepper and the workings of the pepper trade in the Hapsburg period must necessarily be described in provisional terms only.

Piper nigrum, the raison d'être of the old Portuguese India trade on which so much cost and energy was expended for over a century and a half, is a large, vine-like climber, indigenous to the forests of southwest India. The leaves are "smaller than an orange leaf, green and sharp-pointed, burning a little almost like betel."[7] The fruit grows in clusters of small green berries which, on ripening, become "ruby red and transparent cleare."[8] Inside the berry is a soft core surrounded by pulp. Black pepper, the variety usually exported by the Portuguese, is made from the whole fruit, crushed and dried in the sun; white pepper comes from the dried core only, after removal of the skin and pulp. The pepper plant requires a warm, wet climate with an annual rainfall of 100 inches or more, and no frosts. These conditions prevail along the whole littoral of western India from Goa southwards, especially on and near the foothills of the Western Ghats, and it was in Malabar and Kanara that most of the pepper orchards were situated.

A typical pepper orchard consisted of a small strip or plot of land usually located in a hollow or along the line of a depression where moisture and shade were available in abundance.[9] Surrounded by forest, the orchard was first planted with trees such as areca-nuts, mangoes, or betel-palms "orderly sett in ranckes" some six to eight feet apart, on which to train the pepper plants—or else suitable existing trees were retained for this purpose. The farmer planted his cuttings in leaf moulds at the feet of these trees at the onset of the wet monsoon season in June. With the aid of careful watering, and manuring with leaves and brush, the young pepper shoot then grew up the supporting tree "clasping, twyning and fastning it selff theron round about as the ivy doth the oake or other trees with us," and after three years' growth began to bear fruit.[10] The buds appeared on the mature pepper plant in March, and the flowers came out in May. By December the berries began to change color, indicating readiness for harvesting, but for best results they needed to be left on the vines until the middle of that month. Since the berries were delicate and damaged easily, the harvesting had to be done carefully. After they had been dried for a few days in the sun—which shrivelled them and turned them black—the berries were stored for about a month, and were then considered ready for collection by the buyer. The mature pepper plant bore fruit for ten to twelve successive years, after which it was replaced by a new cutting and the whole cycle repeated.[11]

It is likely that few of those who cultivated these pepper orchards in

Malabar and Kanara in the early seventeenth century owned the land which they worked. Most agricultural land probably belonged, as it apparently did under Dutch ascendancy in the late seventeenth and early eighteenth centuries, to landlords (*janmkars*) of three kinds—secular rulers (rajahs), Hindu temples and their attendant Brahmins, and members of the upper Nayar sub-castes.[12] Normally these landlords let out their land to small tenant farmers. Unfortunately there appears to be no contemporary account of the types of tenancy prevailing in Portuguese times. However, the traditional tenancy system of Malabar as described for nineteenth-century British India suggests the kinds of conditions that most probably existed in this earlier period also.

Under the British *Raj* of the last century two classes of tenant were recognized in Malabar. The conditions for the group known as *kanamkaran* varied somewhat, but usually an initial down payment to the landlord was required, and thereafter a one-third share of the annual crop. In return this tenant was entitled to permanent occupancy, though without the right to alienate. The kanamkarans were sometimes members of the lower Nayar sub-castes but were more likely to be *Tiyan* (the low-caste aboriginal peasantry or "toddy drawers"), or Moplahs. By the early nineteenth century kanamkarans held most of the agricultural land in Malabar, which they either farmed themselves with the aid of landless laborers, or sublet to small peasants.[13] The second class of tenant, the *pattamkaran,* had no permanent right of occupancy. He paid over two-thirds of his produce to his landlord and was left with the barest subsistence.[14] He was therefore in a distinctly less favorable position than the kanamkaran.

Although there is no clear evidence for the sixteenth and seventeenth centuries, it seems that the cultivators of the pepper orchards at that time were principally small tenant farmers similar to kanamkarans and pattamkarans of the nineteenth century. It is certainly apparent that the pepper farmers of this earlier period endured a similarly low economic status. Francisco da Costa stated that some pepper was harvested in August and September, before it was fully ripe, and sold inland. "The traders [*mercadores*] are poor, and hunger forces them to do this, despite the fact that they sell this immature pepper at a much lower price than it is worth when ripe."[15] Whether by the term "traders" he meant the cultivators who sold the pepper, or the middlemen who bought it in the first instance, is not clear—in either case it is obvious that the peasant producer sold at disadvantageous prices. When the Dutch inherited the position of dominant European power on the Malabar coast in the later seventeenth century, the pepper producer was still referred to as a small peasant farmer (*kleine landman, arme landman*) which again suggests that his status was essentially similar to that prevailing in the nineteenth century.[16]

The economic subordination of the pepper cultivators is also suggested by the fact that they were frequently forced to pledge their pepper in advance to middlemen. In the early sixteenth century the middlemen would each year acquire the next season's crop either for a cash advance or "in exchange for cotton clothes and other goods which they keep at the seaports."[17] Francis Buchanan, a British official, described a similar debt system in Malabar around 1800, when Indian traders went to the pepper orchards annually between June and July and advanced payment to the farmers, who contracted to deliver their pepper at a specified place in January or February.[18] The advance payments were often made in cloth or other goods, but more usually in *fanams*, the local currency. The price paid and the proportion of the total value actually advanced varied, but if the current need of the cultivator was urgent he got no more than two-thirds of the real value of his pepper. Moreover, if six to eight months later the grower was unable to deliver the agreed quantity, he was forced to compensate the trader for the portion lacking, at the full Calicut price, which was much higher than the normal purchase price in the growing areas.

If, as seems probable, the pepper farmers of southwest India in the sixteenth and early seventeenth centuries were similarly treated, then their lot was hardly enviable, although probably no worse than that of most other Indian peasant farmers of the day. There was certainly profit to be wrung from the pepper trade—but it can be assumed with confidence that only an insignificant share found its way into the hands of the actual tillers of the orchards on the foothills of the Western Ghats.

The middlemen who bought the dried pepper from the cultivators transported it by cart, ox or riverboat to the markets inland or on the coast, and sold it there to distributors and exporters. Most middlemen were independent operators, but some acted as agents for wealthy coastal merchants such as Ramaqueny, a Banyan who provided pepper for the Portuguese India Company in the early 1630's. It is impossible to determine how profitable this middleman's role was.[19]

Although the prices offered by Portuguese buyers on the coast in the sixteenth and seventeenth centuries are known in some instances, there appears to be no corresponding information on the rates at which the middleman paid the grower, and thus no way of calculating gross profit. Buchanan believed that the middlemen in the early nineteenth century "fleeced" the cultivators and made 60-80 per cent on their deals, although it is not clear whether this represents gross or net profit.[20] However, it should not be assumed that profits in the seventeenth century were necessarily always so large. Political instability, periodic famine, difficult and dangerous travel conditions, and a general tendency towards inflation probably caused considerable fluctuations in profits from year to year. Moreover, the Portuguese had a notorious reputation for paying low prices for their pepper. Thus in 1566 the then governor at Goa, Dom Antão de Noronha, informed the crown that the Portuguese, outbid by Moslem merchants from Arabia, would have been unable to buy any Kanara pepper at all but for the armed guards and patrol vessels they maintained against their rivals.[21] In 1634 the Jesuits at Quilon intervened on behalf of pepper-traders whom they alleged were being robbed and cheated by the low prices offered at the Portuguese weighing house. The Fathers contended that the seller should receive no less than the agreed contract price of twenty *patacas* a *bahar,* and some of the stock offered was held back in their college, awaiting resolution of the matter at Goa.[22] The fact that middlemen sometimes resorted to the risky practice of adulterating their pepper also suggests that profit margins may sometimes have been small.

Those traders who served as middlemen in the pepper trade were apparently a heterogeneous group of Hindus, Moslems and Christians. In the early sixteenth century Duarte Barbosa particularly singled out the Hindu caste of *Vyapari Nayars,* dealers in all kinds of merchandise both on the coast and inland, who "gather to themselves all the pepper and ginger from the Nayres and husbandmen." On the other hand, the Dutchman Baldaeus, writing in the third quarter of the seventeenth century, placed more emphasis on the Moslems. He mentioned the "Moors and other merchants" of the Carnatic and Bijapur, who "use to fetch it [the pepper] in considerable quantitys."[23] Both these writers were speaking of the pepper trade of India in general rather than that of the Portuguese in particular—and the Portuguese preferred, when possible, to deal with Christian middlemen.

In practice, however, only a modest fraction of the pepper produced in southwest India and distributed by the various middlemen was bought by the Portuguese. In the entire producing area from Onor in the north to Travancore in the south, the minimum annual pepper production in about the first decade of the seventeenth century was estimated at 100,000 bahars (258,000 quintals), of which not more than 20,000-30,000 quintals were despatched on the carracks to Lisbon—a mere 10 per cent of the total.[24] Most of the remaining 90 per cent was either consumed locally or exported overland to other parts of India. Pepper was taken in considerable quantity by convoys of oxen to the Moghul territory to the north, and northeast across the subcontinent to Bengal. It was also conveyed to the Coromandel coast of southeast India, where some of it was bought by Portuguese country traders who shipped it for resale in Bengali ports, although all trade in pepper was claimed by the Portuguese crown as a strict royal monopoly. Other consignments were brought overland to the Dutch factory at Pulicat and to the Danes at Tranquebar, while pepper was also conveyed, in defiance of the Portuguese monopoly, to the Red Sea and Persian Gulf ports by Malabari and Arab traders.[25]

The pepper bought by the Portuguese for export to Europe was acquired from middlemen at weighing houses and factories in Malabar and Kanara. In Malabar the Portuguese bought from or through local Christians, sometimes utilizing the influence of the church authorities to see that supplies were forthcoming. Probably the necessary contacts were established in the early 1520's when Mar Jacob, bishop of the Saint Thomas Christians, wrote to the king of Portugal that his followers would in future supply no pepper to Moslem traders, but only to the Portuguese.[26] At the end of the decade the Portuguese interpreter in Cochin was affirming that "all the pepper" of the region was in the hands of Christians, and Luís da Costa confirmed at the beginning of the seventeenth century that it was a well-known fact that Saint Thomas Christians brought in most of the pepper to Cochin, and always had done. Even in the difficult years 1603-1605, when pepper was scarce on the Malabar coast, Christian merchants were persuaded, largely through the efforts of the bishop and archdeacon of Angomale, to keep the Portuguese factory at Cochin supplied.[27] In Kanara, where there was no significant native Christian community, conditions were somewhat different. Here the Portuguese practice was to arrange contracts with individual merchants or syndicates, whether Portuguese,

Eurasian or Indian, Christian or Hindu. For example, in 1602 a certain António Mendes de Tomar and António Fernandes de Sampaio contracted to supply the Portuguese in Kanara with pepper, and in 1603 an Indian goldsmith and his nephew agreed to supply 1,500 quintals at Onor and Barcelore. Later in the same decade the Portuguese were buying pepper at Onor on contract from some Brahmins.[28] Finally, with the rise of Venkatapa Nayak (1602-1629) the Portuguese bought most of their Kanara supplies on contract from the nayaks of Ikkeri, though by force of circumstances rather than by preference. These various contractors acquired their stocks either through sending their own agents to buy in the producing areas, or indirectly from the lesser middlemen.

Since the pepper was harvested in southwest India in late December or January, and required about a month's drying in the sun before it was considered suitable for sale, supplies were not normally bought by the Portuguese factors before March or late February at the earliest. Usually the buying by the Portuguese was done in the two or three months from the beginning of March onwards, and was supposedly completed by the end of May, since thereafter supplies dwindled. It was often possible to purchase further consignments in November or December if those acquired earlier were insufficient, but buying so late usually meant poorer quality.[29] Nevertheless, the Portuguese were often forced to resort to this expedient in order to fill carracks that had to clear Goa for Europe well before the following March. This happened in 1629 when most of the pepper bought by the Portuguese India Company at Cochin was acquired between November 29 and December 30.[30]

The weighing, assessing and handing-over of the pepper took place at a kind of closed market attended by Portuguese trade officials, officials representing the local ruler, and the traders selling the pepper. Each of the principal exporting towns had a weighing house set aside for these transactions, equipped with the necessary scales and other facilities. These buildings were invariably at points of easy access to river transport. In Malabar there were weighing houses at both Cochin and Quilon. That at Cochin was "a house built on the river 30 paces from the beach" or, as Tavernier put it some years later, "a large store surrounded by the sea."[31] It was clearly marked on contemporary plans of the city, and its location suggests that it may originally have been part of the fortifications. By the 1630's, however, it had long been in use for the pepper weighing, but the ravages of time had brought it to such a leaky and dilapidated condition that it was near the point of collapse, and a palm-frond shelter had had to be made to serve as a temporary substitute. Apparently the old building was later repaired, for the Portuguese used it as a defense point during the Dutch siege of 1653. In Quilon the weighing house was similarly located near the water, probably in the vicinity of the Portuguese fort.[32]

In Kanara an important weighing house was located on the Sharavati River near Gersoppa, a town in the pepper country up-river from the Portuguese factory at Onor. This building, like the others, was looked after by a resident guard or caretaker who permitted travellers to lodge there out of season. The Italian Pietro della Valle, who spent the night of October 21, 1623, there, described it as "a place cover'd with a roof amongst certain trees, where many are wont to lodge, and where the pepper is weigh'd and contracted for when the Portugals come to fetch it."[33] Other Portuguese weighing houses were in Onor, Barcelore and Mangalore, and in Goa itself.

Procedure at the weighing sessions was probably roughly the same in all these centres.[34] At Cochin the Portuguese *vedor da fazenda* and a representative of the rajah known as the *regidor-mor* presided jointly, each seated on an upright chair. Also present were the Portuguese factor, scribe and weight inspector (*juiz do peso*), and the rajah's clerks, who sat all together on a bench. The weigher (*pesador*), a Hindu in the employ of the Cochinese ruler, placed the pepper in half-bahar portions in three or four open sacks on one side of the scales, and three weights on the other side together with an equal number of empty sacks to make a precise balance. The scribe entered the date, weight and name of seller in his account book, and the pepper was then put aside on straw mats. There it was carefully inspected by the weight inspector and the scribe to ensure that it was in an acceptable condition, and not damp or contaminated.

This inspection was important, for the unwary buyer could easily be duped. Sometimes the middlemen offered immature pepper for sale in November, claiming it had come from an early harvest. Actually, genuine early ripening was rare and pepper offered so early in the season was usually prematurely picked and improperly dried. Nevertheless, inability or failure to buy pepper in the favorable March-May period sometimes forced the Portuguese to do so in November or December, so as to fill the carracks that had to clear Indian ports for the run to Europe well before the fully matured pepper from March onwards became available.

Portuguese officials at the buying centres also had to be on the lookout for traders who mixed a small quantity of mature pepper with the immature, in an effort to pass it all off as fully dried. Berries improperly dried at the time of purchase were liable to suffer substantial losses in weight by the time they reached Europe, thus causing serious reductions in profit. Improperly dried pepper could be recognized by its distinctive odor, its relative softness and the damp appearance within the husk when broken open. Some middlemen adulterated their pepper with ash before it reached the weighing houses. The ash was washed into the wrinkles of the pepper berries with water and might remain hidden there for a month or so, until the heat began to draw it out. The bluff could be recognized at the weighing by chewing a few berries and seeing if they tasted of ash. Of course, sharp practice on the part of suppliers in India was not the only cause of subsequent losses in weight. These could result from the dishonesty of corrupt Portuguese officials who might handle consignments before they arrived at the Casa da Índia in Lisbon. Moreover, even if the pepper was mature and unadulterated, and the officials handling it were honest and competent, some losses still occurred, if only because the friable quality of pepper made wastage inevitable during the processes of sacking, loading and transporting by sea.[35] Although this could be minimized by careful handling, it could not be completely eliminated.

By the beginning of the second half of the sixteenth century the Portuguese were experiencing increasing rates of weight loss on their pepper. A major cause was their attempt to hold down pepper prices in the buying areas at a time of general inflation, thus forcing suppliers in India to offer low quality or adulterated stock in an effort to cut their own costs.[36] However, the position began to improve after about 1585, when the Portuguese vedor da fazenda at Cochin, Nicolão Pedro Coceno, agreed with the rajah to pay the middlemen a supplement of 12 per cent on the old price for pepper that was genuinely mature and unadulterated.[37] Inspection procedures were probably tightened at the same time, for at Cochin in Francisco da Costa's day, a half-bahar sample from each supplier was normally left out on straw mats for a further drying period after initial weighing, then returned to its sacks and reweighed. Any weight loss noted on the second weighing was subtracted not only from the sample load, but from the entire consignment of that particular supplier, at the same rate. A careful watch had to be kept when the sample was drying, as the trader might try to add extra pepper to bring it up to scratch at the reweighing, thus avoiding a reduction in the price of his total consignment. These precautions were apparently quite successful in persuading traders to offer clean pepper at Cochin, and Francisco da Costa estimated in the early seventeenth century that the total loss of weight incurred by pepper in transit between the Malabar and Kanara factories and Goa averaged only about 3 per cent.[38]

On completion of the weighing and reweighing the pepper was packed in gunnysacks which were sealed up with coconut-fibre twine by Indian binders. The sacks were then transported down-river by boat to the factory where they were kept padlocked and bolted in special stores until the arrival of the annual fleet of coastal craft. These vessels transported the pepper to Goa where it was reweighed and then loaded aboard the India carracks for trans-shipment to Lisbon.

The weighing, buying and transporting of pepper at or to Cochin, Quilon and other points of export involved the Portuguese in a variety of expenditures over and above the cost of the pepper itself. There were the salaries, perquisites and expenses of Portuguese factors and other trade officials, the wages and benefits paid to guards, coolies and various other employees, general maintenance of facilities, and freight costs. To these should be added the pensions, gifts and bribes given to rajahs, petty chiefs and native officials, whose goodwill was considered desirable to ensure the smooth flowing of the pepper into the weighing houses and the factories.

The salaries paid to Portuguese trade officials accounted for a relatively trivial portion of total funds. Cochin, with the largest buying centre, was supposed to run a basic staff of one factor, two scribes and a weight inspector. The factor's salary was set at 716 xerafins, while the clerks and *juiz do peso* each received 133 xerafins.[39] In smaller centres such as Onor, Barcelore and Mangalore there were no factory scribes or weight inspectors, and the weighing and accounting in these places was handled by the factor himself. On the other hand Cochin supported, in addition to the trade officials mentioned, a vedor da fazenda and a treasury secretary who played an important role at the weighing. Some of these officials also enjoyed a variety of perquisites. From the load of pepper brought by each river-craft to the weighing house at Cochin, half a bahar was given to the juiz do peso and the two factory scribes, who split it three ways. Since the standard load of a river-craft was reckoned to be thirty bahars, the portion taken by these officials amounted to 1.7 per cent of the total. The same three officials also shared among them one *vintém* for every bahar weighed in the weighing house and were given a further 6 *réis* per bahar each when the pepper was weighed again at the time of its trans-shipment.[40] The costs of these last two perquisites were born respectively by the rajah of Cochin and the Portuguese buyer, while the half-bahar of pepper per river-craft was provided at the expense of the traders or middlemen.

The kinds of expenses the Portuguese trade officials charged up can be seen in the annual accounts submitted to the Lisbon board of the India Company in 1629. That year the company's factor for Kanara claimed to have spent 250 xerafins on an unsuccessful journey to Onor in search of pepper. He claimed a further 5 xerafins for submitting protests to the authorities at Onor when the pepper failed to appear, while other claims included costs for the hire of boats to take him upstream to Gersoppa on three separate occasions. During the same year the accounts for Cochin include charges for such items as the hire of palenquins for going to Upper Cochin "to speak to the king about the pepper coming to the weighing."[41]

The Portuguese factories in Malabar and Kanara also employed various local people—some, such as the caretakers of the weighing houses, on a year-round basis, but most as seasonal laborers. The caretaker at Cochin enjoyed the right to live in the weighing house, together with a small perquisite from the load of each river-craft that reached it. He received no wages and no other perquisites apart from the sweepings of the traders' pepper mats in return for guarding their pepper and providing them with water and lighting at night. However, wages were paid to the numerous coolie laborers who were required from time to time in all the factories, for such tasks as carrying the sacks of pepper to and from the boats, the stores, and the weighing houses. The largest single item in the operating costs of the Cochin factory in 1629, apart from the salaries of the Portuguese officials, was 477 xerafins for teams of coolies who worked at the drying and weighing of the pepper. At Goa in the same year 1,362 coolies worked eleven days for the Portuguese India Company, hastily loading the carracks for Lisbon, while the Kanara factor, Diogo Vaz Pereira, took 25 laborers with him to Gersoppa to work on shifting the pepper, providing them with rice and fish at company expense. The wages paid these coolies were very low. The men who loaded the Kanara pepper for Goa in 1629, presumably at Onor, received 48 réis, or less than one-sixth of a xerafim each, for an unspecified number of days' work. Yet a standard bale of rice cost 390 réis in Kanara that year, and 570 réis in Goa, and these rates were cheap compared to the astronomical sums charged in the famine years of 1630 and 1631.[42]

Additional employees were needed for making, repairing and sewing up the gunnysacks in which to pack the pepper. They received 7 réis (less than one-fortieth of a xerafim) per sack. Rather better off were the headmen, who supervised the coolies and were paid double the normal wage, and the *pattamars,* who took messages overland on foot from one factory to another. A pattamar sent from Cochin to Quilon in 1629 was paid 3 xerafins for his services. In a quite different category were the better paid "trusty natives" who acted as interpreters to the factor, or in other responsible capacities. Three Brahmins who accompanied Diogo Vaz Pereira on his journeys in 1629 "to help receive the pepper and to go as guards and come back in the pepper boats" were given 40 xerafins to divide among them.[43] In the buying season factors also tried to enlist the aid of experienced Portuguese residents who had local knowledge and could assist in negotiations. Guards were needed for the specie and pepper, especially when in transit to or from a weighing house some distance inland, such as Gersoppa. Local Portuguese or Eurasians were usually employed for this purpose—but in small numbers, and for payment which varied quite widely according to locality, time and service rendered.[44]

Each of the Portuguese factories in Kanara and Malabar required annual supplies of cloth, thread and needles for making gunnysacks, straw mats and baskets for drying and carrying pepper, and candles and oil for lighting. It was also necessary to maintain the weighing houses, but it appears that as little as possible was spent for this purpose and that makeshift arrangements were often tolerated. Bocarro's comments on the dilapidated condition of the weighing house at Cochin have already been noted. In 1629 the weighing house at Goa was little better, for it consisted of a temporary palm-frond shack, the building normally used for the purpose having been taken over to house artillery from the carrack *São Gonçalo* and from the royal galleons.[45] Nevertheless, the most essential maintenance work had to be done to keep the pepper trade functioning, and several hundred xerafins were spent each year on such necessities as keeping the scales in good repair, and doors, bolts and padlocks properly secured.

The cost of transporting the pepper by river and by sea on the various stages of its journey to Goa had also to be taken into account. At Cochin in 1629 the Portuguese paid 1½ *tangas* per journey for the river-craft which carried the pepper the relatively short distance from the casa do peso to the factory, and in Kanara they spent over 155 xerafins for the hire of boats to bring the pepper downstream from Gersoppa to Onor.[46] These expenses were relatively trivial—but more costly was the freighting of pepper on the fast coastal vessels from the various factories to Goa, there to be loaded on the carracks for shipment to Europe. These cargo vessels—mostly galliots propelled by a combination of sails and oars—were despatched each year by the viceroy at Goa. They sailed in convoys under armed escort, as protection against Malabar corsairs. Reaching southerly ports like Cochin soon after the fading of the wet monsoon season, the galliots were required to be back in Goa in time to load the carracks for Europe in January or early February. The cost of freighting pepper on these vessels usually amounted to a few thousand xerafins, as in 1629 when the Portuguese India Company paid 1,117 xerafins for the despatch of its pepper to Goa from Cochin and 3,009 xerafins from Quilon.[47]

These various charges which the Portuguese had to meet in the handling of their pepper between the time of its purchase and the time of its despatch to Lisbon in the holds of the carracks added relatively little to their outlay. Under the regime of the Portuguese India Company, they effected an increase of about 10 per cent over the cost price of the pepper. However, if duties, pensions and other payments made to the rulers of the pepper-producing and exporting areas are also added, the increase jumps to perhaps 25 per cent, indicating the importance of these local authorities in the Portuguese trade system.

Although the Portuguese pepper supplies from Malabar and Kanara were bought through middlemen, the procedure itself was not exclusively commercial. In these areas the Portuguese tried to organize their pepper trade on the basis of political agreements with the petty Indian states.[48] In particular, they preferred that the local ruler take responsibility for the safe passage of the pepper, that he guarantee that pepper not be sold to other buyers, and sometimes that he arrange for supplies to be made available to the Portuguese factories. For services of this kind, and also simply to retain the goodwill of the local ruler, the Portuguese were obliged to make various payments to him, and some of these had long been established custom by the Hapsburg period.

Customs duties, which the Portuguese had paid in Malabar since the beginning of the sixteenth century, were the most important of these payments. In Cochin the rate given to the rajah had originally been fixed at half a cruzado a bahar, which amounted to an impost of 6.25 per cent on the price of 8 cruzado a bahar the Portuguese were then paying for their pepper. However, in 1569 vedor da fazenda Vasco Lourenço made a new agreement with the rajah of Cochin whereby the Portuguese undertook to raise the duties they paid the Cochinese ruler to two gold *São Tomés* (the equivalent of three xerafins)—thus doubling the previous rate to 12.5 per cent. In return for this increase, the rajah undertook to protect the river-craft in their journeys down-river from the pepper-producing areas and to pay the various annual pensions which the Portuguese customarily gave the petty rulers of the pepper country. This arrangement appears to have remained in effect during the Hapsburg years.[49] In Quilon the Portuguese had originally paid customs duties at the same rate as in Cochin, but in time the arrangements in the two settlements diverged somewhat. In the early seventeenth century the Quilon duties were apparently shared between the rani and her more powerful neighbor and overlord, the rajah of Travancore, who in turn shared out some of the proceeds among their leading officials. By the late 1620's, however, the Portuguese appear to have paid some of the duties directly to the petty rulers of neighboring pepper districts, at a rate calculated to be about 8 per cent.[50]

In the Kanara factories at Onor, Barcelore and Mangalore, it was not the Portuguese who paid the customs duties, but rather the traders who supplied the pepper. However, the Portuguese gained no real price advantage from this arrangement, and invariably paid more for the good quality Kanara pepper than for Malabar pepper. In Francisco da Costa's day a light quintal of Onor pepper sold for 13 xerafins 20 réis; at Cochin the same amount of pepper sold for 9 xerafins 2 tangas 26 réis, or 40 per cent less.[51] Moreover, the running costs of the Portuguese establishments in Kanara tended to be higher than those in Cochin or Quilon, since the Kanara factors were obliged to do more travelling in the search for pepper than was the practice with their colleagues to the south. Finally, from the 1610's the Portuguese were obliged to buy much of their pepper in Kanara on expensive contracts from Vira Bhadra Nayak of Ikkeri, whereas in Cochin and Quilon they were tied by no such agreements.

In addition to the customs duties levied by the rulers of Malabar, the Portuguese paid, by tradition and policy, a whole series of gifts, perquisites and pensions to the various princes and officials of the pepper-producing and exporting areas "in order that they should favour and not hinder the pepper passing through their territories."[52] Probably the largest payment, and perhaps the only one that was not deliberately suppressed at some time or other in the sixteenth or early seventeenth centuries, was the rajah of Cochin's *copa*. This was a special annual pension of coin to the value of ten marks of gold, normally handed over with pomp and ceremony after the pepper had been loaded aboard the ships for Goa and Europe.[53] Smaller pensions known as *tenças* were paid to many of the neighboring kings (*reis vizinhos*) of the Malabar pepper lands, who supplied pepper for the carracks. Pension lists for the years 1554, 1564 and 1607, which agree fairly closely, give the identities of the leading recipients. The most important were the pepper king and his mother, who head all three lists. There were also the rulers of Purakkad, Mangatti, Parur and Diamper, and a number of others whose titles are too confused in the Portuguese to be identifiable.[54]

The pensions of the neighboring kings had a somewhat chequered history. In the early 1540's the viceregal government attempted to suppress them as an economy measure, and this led to a serious "discontent."[55] Led by the pepper king, most of the reis vizinhos attempted to switch loyalties from the pro-Portuguese rajah of Cochin to the hostile samorin of Calicut, and a long and bitter war ensued, with much fighting on the channels and backwaters of the south Malabar coast. Following this, the Portuguese restored the pensions, but after 1569 they ceased to pay the kings directly, arranging for this to be done instead through the local overlord, the rajah of Cochin. By the early seventeenth century the system was apparently in need of further review, for Francisco da Costa recommended that only the pepper king and another ruler not on the earlier pension lists, the rajah of Turuguly, should be paid regularly, since they alone were making regular deliveries. Payments should not be made to the ruler of Purakkad, he added, "because in all my time I have not seen a single boat from his lands at this weighing point"; likewise from Mangatti, Parur and several other localities arrivals had become very infrequent. Da Costa also warned that local rulers were permitting pepper to be diverted into the interior east of the Ghats because "the traders of the backlands, knowing that we give them a pension of 100 cruzados, give them 101, and for this one they [the rulers] forget the obligation that they owe."[56]

Finally, to various Indian rulers, and to officials, merchants and laborers who attended at the purchase proceedings, the Portuguese distributed gifts in kind. These usually included supplies of betelnut and a square meal for those bringing the pepper to the weighing houses, and such presents as coats, jackets and caps to persons considered more important or deserving. In addition, important Indians were sometimes paid or bribed for special services rendered—such as Vitoba Sinai, Vira Bhadra Nayak's ambassador, who in 1629 received 456 xerafins for seeing that the pepper bought on contract in Kanara that year was duly delivered.[57]

There is little doubt that the advantages the Portuguese gained in terms of cooperation from local Indian authorities were well worth the few hundred xerafins expended each year on pensions and presents, and the larger sums surrendered as customs duties. In fact, by the early seventeenth century the Portuguese were expending much less for these benefits than they had done in the sixteenth century, since all the traditional duties and perquisites had sunk in real value owing to the general inflation. The most important item, the Cochin customs duty on pepper, had been revised once only, in 1569; other payments had also failed to keep pace with the times. The extent of the decline in value is illustrated by the fact that the pensions, which in the mid-sixteenth century had been worth the equivalent of 644 quintals of pepper, were worth only 288 quintals by the early seventeenth century.[58] Moreover, there must also have been a decline in the amounts payable as customs duty in the early seventeenth century, owing to the smaller quantities of pepper the Portuguese were then exporting.

On arrival at Goa, the pepper from the Malabar and Kanara ports was reweighed at the casa do peso and loaded aboard Indiamen for the voyage to Portugal. The unit employed at the reweighing was the light quintal of 128 *arráteis* or pounds of fourteen ounces, as used in the Casa da Índia at Lisbon, rather than the heavy quintal of 128 arráteis of sixteen ounces each, normal in the factories of India. The weight in light quintals as determined at the reweighing in Goa was the version used in the ships' manifests.[59] The pepper was supposed to be loaded on board the carracks as soon as possible after its arrival in Goa, in order to facilitate a prompt sailing for Europe. In practice delays in loading were frequent, resulting in a sudden rush in January and early February, and in mutual complaints and recriminations among the authorities at Goa and Lisbon, the ships' officers and, in the early 1630's, the directors of the India Company. These resulted from failure to complete the purchasing of pepper and other commodities sufficiently early—which in turn occurred because the Portuguese factors in Malabar and Kanara had not been provided with enough money for the preceding buying season.

Pepper was always the first commodity to be stowed on board the carracks. Detailed standing orders were designed to ensure that this was accomplished with "great care and diligent watch." The vessel's ballast was first covered with deal boarding, and in the two levels or decks immediately above it about thirty compartments called *paioes* were divided off, normally by region of origin. On completion of the lading, these compartments were sealed "verie close, with ocam and pitch," numbered, and marked with the weight of the pepper[60] The treasurer-general, the carrack's master and various officials and clerks of the treasury department in Goa were required to be present at the final weighing, the lading, and the tarring of the paioes. Small samples of pepper from each region of origin, or from each paiol in the carrack, were to be placed aside in a sealed container. Weight losses noted in any particular portion of the cargo on arrival in Lisbon could then be compared with those incurred by the corresponding sample.[61]

Despite these precautions, dishonest or careless officials in Portuguese India or on the voyage home could and did cause serious losses. These were so great on the pepper cargoes of 1610 that Garcia de Melo, the vedor da fazenda at Cochin, was ordered arrested and sent back to Lisbon for punishment.[62] This action failed to improve matters, for in 1611 2,000 quintals of pepper worth 70,000 cruzados were lost on the flagship of the homebound fleet, allegedly because the cargo had been laid straight over the ballast in order to make compartments available for sale to private individuals.[63]

The weight loss incurred on pepper between its enshipment at Goa and arrival in Lisbon varied considerably from year to year. Rates in the early and late sixteenth century were in the region of 6 to 12 per cent, but at mid-century they appear at times to have reached as high as 40 per cent. The standing orders in force at the Casa da Índia in the 1630's note—and deplore—that serious losses on pepper consignments reaching Lisbon had become "almost customary." Nevertheless, the record could not have been consistently bad in the early decades of the seventeenth century for the Portuguese India Company's carrack *Santíssimo Sacramento* suffered losses of about 2 per cent only, on the Goa-Lisbon voyage in 1630.[64]

Even without human greed and laxness, accidents sometimes intervened to prevent cargoes from reaching Lisbon safely. At the worst, pepper-laden carracks and galleons might fail to complete their journeys at all, instead ending their days as wrecks on the coasts of southeast Africa or on the reefs and islands of the Indian Ocean, or as prizes in the hands of Dutch, English and Algerian enemies in the Atlantic. Shipwreck was the fate of the Portuguese India Company carracks *São Gonçalo* in 1630 and *Santo Inácio de Loiola* in 1632, and of an alarming proportion of Portuguese Indiamen in the late sixteenth and early seventeenth centuries. Lesser mishaps on the voyage home could also result in loss. Pepper and other spices were particularly susceptible to damage by contact with seawater, a danger compounded by the fact that serious cases of such contamination could result in the formation of lethal fumes in the ship's holds.[65] Also, storm conditions might necessitate jettisoning some of the cargo, and in such an emergency the master—as a last resort and in the presence of suitable witnesses—was permitted to open the pepper compartments. In this event regulations demanded a full, written report on the incident when and if the vessel eventually reached port.[66]

On the arrival of a pepper-laden carrack in Portuguese waters the possibility of further losses through smuggling or theft increased, and precautions had to be taken accordingly. Only the most essential contacts were permitted between persons in a returning Indiaman and those on board escorts and other vessels it might encounter as it approached home.[67] As soon as the *provedor-mor* of the Casa da Índia knew an India carrack was approaching the bar of the Tagus he was to appoint armed customs officials to board the ship at Lisbon, supervise the discharge of cargo, and search passengers and crew at disembarkation.[68] Unloading goods from the ship for clearance or storage at the Casa da Índia was to be swift, orderly and carried out in daylight.

After private luggage had been cleared, work on unloading the pepper began. Regulations required that the provedor-mor himself attend this procedure and that, before reweighing began, the ship's manifests be carefully studied and collated, and their contents copied into the official casa registry. The reweighing itself was supervised by a weight inspector from the casa, and only casa officials and the ship's master were permitted to attend. As a double check, two scribes recorded the results independently.

When the weighing and registering of the pepper, and the calculation of losses, were complete, the pepper was transferred to carefully labelled compartments in the casa warehouses and neatly stacked away from the potentially damaging influences of heat and moisture. There it awaited inspection and purchase by the wholesalers and their agents, who handled the distribution of pepper to the markets of Portugal and Europe.

4 / Crisis in the Early Seventeenth Century

During the first four decades of the seventeenth century Portuguese trade in Asia—with a few exceptions, such as the Macao-Nagasaki trade—was everywhere declining. In consequence, the revenues of the viceregal government at Goa were severely reduced just when official costs, particularly for defense, were rising to unprecedented levels. Decreasing revenues and rising costs were in fact the two principal components of the Estado da Índia's economic crisis in the later Hapsburg period and made almost inevitable Portugal's elimination as a major power beyond the Cape between the late 1630's and 1650's.

The inability of the viceregal government at Goa to meet its defense costs from the regular revenues of the Estado da Índia was recognized by Lisbon as early as 1589—before the Dutch and English had become serious threats to Portuguese predominance in Asian waters.[1] A quarter of a century later, when these threats had become alarming and expensive realities, François Pyrard, who had lived at Goa from 1608 to 1610 and knew it well, wrote that "the revenue of the [Portuguese] Indies cannot at present be sufficient to pay and maintain the State . . . and the cost is greater than the value."[2] That this situation persisted and in fact worsened is confirmed by repeated reports from the Portuguese administration at Goa in later Hapsburg years. Every new viceroy from the Count of Redondo in 1617 to the Count of Aveiras in 1640 complained vehemently on arrival in India that the Goa treasury was empty, and funds unavailable for the most pressing needs of state. Frequently the Goa administration was unable to meet from its regular local revenues even routine expenditures, let alone extraordinary defense needs.[3]

In the Goa territories themselves decreasing revenues were primarily a product of reduced returns from the customs. In the early seventeenth century between one-half and one-third of regular annual receipts from these territories was being provided by customs dues, but as the Portuguese seaborne trade in Asian waters contracted in both volume and variety, so customs revenue declined—a trend much stressed in the lamentations of contemporary Jeremiahs. In a treatise on Portugal's problems in Asia, written in Lisbon in 1611, the author complained that the loss of customs duties paid on spices in the State of India was making it impossible for the Portuguese to fit out the fleets they needed.[4] A second writer alleged that the Portuguese in Asia were "losing the duties of all the customs houses and the trade and traffic of the sea without which they cannot sustain themselves."[5] Early in the 1620's the German merchant Ferdinand Kron, who served as the Fuggers' and Welsers' agent in Portuguese India between 1587 and 1624, agreed that the Goa customs revenue "is today diminished," while a report submitted to the crown in 1627 explained that revenue from the Goa customs was half what it had been a decade before, and that in consequence what had previously been a substantial surplus on current account had been transformed into a deficit.[6] Somewhat later, in the mid-1630's, the authoritative António Bocarro stated that the profits from many Asian commodities handled by the Portuguese, such as Cambay cloth and indigo, had declined. The volume of trade was shrinking, and he predicted that the yield from customs, which was now variable and uncertain, could only be expected to diminish.[7]

The figures for the Goa customs returns, though fragmentary, give some substance to these reports (Table 1). They suggest that early in the century a gradual decline set in, reaching its lowest point in 1628; by 1634 the decrease from the 1600 figure amounted to about 40 per cent. In fact, the decline in real value was greater, owing to the decreasing purchasing power of the *xerafim,* in a period of inflation.

Table 1. Goa Customs Returns (in xerafins)

Year	Amount
c. 1600	226,666
1610	200,000
1627	150,000
1628	125,000
1629	132,000[a]
1634	130,000

Sources: Luiz de Figueiredo Falcão, *Livro* (Lisbon, 1859), p. 75; Évora, codex cxvi/1-18, f.6; BN Rio, codex 2/2-19, f.325; Évora, codex cxvi/2-3, ff.69-72; LP I, 266.

[a]The slight increase was probably achieved when the new viceroy, the Count of Linhares, forced a tougher bargain with the contractor (Ajuda, codex 51-vii-12, f.50v).

The Goa territories also yielded regular revenues from quitrents on crown lands in Salcete, Bardez, Tissuary and the smaller islands nearby, and from a number of sales taxes which were levied on various commodities ranging from betel to tobacco and contracted out to tax-farmers. Estimates of revenue for the Goa government from all local sources, and of regular expenditure, are known for the years about 1610, 1630 and 1634 (Table 2). While the run is too incomplete to establish a trend, it does suggest that whereas in about 1610 regular revenues could be expected to exceed routine expenditure by a substantial margin, by the 1630's the opposite was the case. Inability to fulfill regular commitments from local income was certainly the topic of frequent vociferous complaints by viceroys and governors in the 1620's and 1630's. In 1627 the Count of Vidigueira informed the crown that regular salary obligations totalling 70,000 xerafins could not be met by the Goa revenues to which they were assigned.[8] This was before the many pressing needs of fleets and fortresses could be even considered, and Vidigueira asked for instructions as to what he should do in these circumstances. Ten years later precisely the same complaint was made, Viceroy Pero da Silva reminding the crown how little money the royal treasury in India contained "because normal expenditure exceeds the revenue of the state."[9]

Had most of the Portuguese administrations in the outlying centres of the empire in Asia been wealthy, or even merely solvent, then the

Table 2. Estimates of Local Revenue and Expenditure for the Viceregal Government at Goa (in xerafins)

Year	Revenue	Expenditure	Balance
c. 1610	439,826	249,654	190,172
1630	321,923	409,232	(- 87,309)
1634	304,346	339,424	(- 35,078)

Sources: Évora, codex cxvi/1-18, ff.6-14v; BN Rio, codex 2/2-19, ff.179-180, 325-326; LP I, 296-300.

Note: Revenues for each year are made up of customs and other *rendas*, and *foros* from the Goa territories. Items from outside sources such as Sri Lankan cinnamon have been excluded. Expenditures consist principally of civil and ecclesiastical salaries, allowances and pensions.

predicament of the viceroys in an era of decreasing prosperity at Goa need not have been so serious. However, unfortunately for the Portuguese most of their subsidiary towns and trading posts were equally short of funds. By 1634 only the northern centres of Bassein and Chaul, and to a lesser extent Mozambique and Damão, were producing significant surpluses, the remaining settlements consuming their incomes locally and often requiring subsidies from Goa. Apart from these four pockets of relative prosperity, and the special case of Macao, all the significant Portuguese centres in Asia were in financial difficulties.[10] Malacca, highly lucrative in the mid-sixteenth century, had become a liability, and in dire need of financial and military assistance from Goa and Portugal. In 1629 conditions were so bad in Malacca, then under pressure from the Dutch, that no revenue at all was forthcoming from the local customs-house and salaries remained unpaid, while the viceregal government sought desperately for money with which to assist the city.[11] Ormuz, which in the 1550's had boasted the most lucrative *alfandega* in the whole of Portuguese Asia, was lost with all its revenues to the Persians and English in 1622.[12] Diu on the coast of Gujarat had been the major port of exit for Cambay cloth and indigo at the start of the seventeenth century, and had yielded 230,000 xerafins a year in customs duties. By the 1620's, however, it had been largely superseded by the Moghul port of Surat, where the English and Dutch had established themselves, and in consequence the customs revenue of the Portuguese city had declined by more than half.[13] "The expenses of this port," wrote António Bocarro of Diu in the early 1630's, "were formerly met from its own income, and many thousands of xerafins were left over and were brought to Goa, as well as donations which the inhabitants sometimes gave to help the state in its expenses." The time had now come, however, when it was necessary to make provision for Diu from Goa, and in 1634 20,000 xerafins had had to be forwarded to Diu from the viceregal treasury.[14] The overall estimates for revenue and expenditure in the outlying centres in 1630 and 1634 show that, despite the efforts of the Count of Linhares, one of the ablest viceroys of the century, Portugal's subsidiary possessions in Asia were running at an overall loss to the government. The loss was particularly large in places directly involved in military operations, such as Muscat and Columbo (Table 3).

Available estimates for the State of India as a whole—that is, total combined estimates for the Goa territories, the outlying centres, and also revenue and expenditure channelled through Goa but emanating from outside sources—present at first glance a somewhat inconclusive picture (Table 4). They seem to indicate a substantial surplus in c. 1610, a deficit in 1630, and another surplus, though relatively small, in 1634. The apparent surpluses in two of these years might seem to cast some doubt on the contention that Portuguese governments in Asia were experiencing an increasingly critical cash shortage in later Hapsburg years. However, a further examination of the data shows that the supposed surpluses of both 1610 and 1634 were illusory. Francisco Pais, the author of the 1610 report, describes the "surplus" that appears in his estimates as constituting money to buy galleons, galleys, artillery, gunpowder and other war materials.[15] In other words, it is really part of a defense allocation, since no provision for this vital purpose was included under regular state expenditure. Similarly, in Resende's 1634 estimates there is no allowance for extra-

Table 3. Government Revenue and Expenditure for the Subsidiary Centres of the State of India in 1630 and 1634 (in xerafins)

	Revenue		Expenditure	
Centre	1630	1634	1630	1634
Mozambique	42,188	42,500	39,104	35,291
Muscat	53,903	67,125	78,724	120,952
Diu	101,028	45,094	94,416	81,527
Damão	49,103	53,519	47,129	35,537
Bassein	115,097	122,469	107,512	58,895
Chaul	23,575	53,687	19,300	17,271
Jaffna	28,363	23,190	29,112	21,797
Columbo		40,774		74,241
Malacca	38,037	37,561	54,387	70,092
Mombasa	17,397	7,717	18,763	16,097
Cochin	2,332		36,017	15,487
Others	9,137	21,256	25,578	26,679
TOTAL	480,160	514,892	524,464	573,866

Sources: BN Rio, codex 2/2-19, ff.179-180v; BNL FG, codex 1783, ff. 109-113.

Note: All centres whose revenues or expenditures were not calculated as exceeding 10,000 xerafins on at least one occasion in the two years have been grouped under "others."

Table 4. Estimates of Total Revenue and Expenditure for the State of India (in xerafins)

Year	Revenue	Expenditure	Balance
c.1610	1,301,986	856,594	445,392
1630	801,992	960,172	(-158,180)
1634	1,185,264	1,115,455	69,809

Sources: BN Rio, codex 2/2-19, ff.353v-354, and codex 2/2-19, ff.179v, 180v; BNL FG, codex 1783, ff.109-111v.

ordinary outlays, military or otherwise, although such outlays would certainly have ensured a substantial deficit in practice.[16] The 1634 figures are also inflated in that they include under revenue several important items which in reality could not be relied upon to provide the sums specified. Among these were 100,000 xerafins optimistically included as the crown's share of the Sri Lanka cinnamon, whereas very little of this commodity had been forthcoming since the uprising against Constantino de Sá in 1630 and the ensuing wars. Likewise the large sums anticipated from the Macao-Nagasaki voyages were far from secure since the success of these ventures was unpredictable in the troubled 1630's, and the trade was to cease entirely in 1638.[17]

Urgent military requirements undoubtedly absorbed any funds left over after routine expenditures had been attended to in later Hapsburg Portuguese Asia. This fact is well illustrated by the budgetary problems of 1630 when the viceregal government had to find an extra 180,000 xerafins for the Goa *terço* (regular infantry regiment), 50,000 xerafins as a military subsidy for Rui Freire de Andrade's forces based at Muscat, and lesser sums as military aid for Malacca, Sri Lanka, Mozambique and São Tomé.[18] Later in the year three disasters to Portuguese arms occurred—the drowning of the able naval commander Nuno Álvares Botelho in a Sumatran river and the loss shortly afterwards of much of his fleet in a storm, the fall of the fortress of Mombasa to a coup by the local sultan, and the defeat suffered by Constantino de Sá in Sri Lanka. It is clear from Viceroy Linhares' diaries for 1630 and 1631 that these developments occasioned further heavy demands on the Goa government's seriously depleted treasury.[19]

The Portuguese authorities in both Europe and India well understood that the viceregal government at Goa was in acute financial difficulties in the later Hapsburg years. However, finding and applying effective remedies proved extremely difficult, though a number of measures were attempted, or at least considered. Among these were the introduction of economies in non-essential areas and a revision of the system of priority claims on the limited funds by then available to the Estado da Índia. The Lisbon authorities seem to have issued only general admonitions to Goa to cut back government spending where possible. The details were left to viceroys and their councils, who were naturally reluctant to enact serious stringency measures which to be effective would certainly have lightened their own pockets. When Viceroy Linhares attempted to work out an economization plan with leading church and civilian officeholders at Goa in March 1630, he began by proposing reductions in his own viceregal establishment which would have saved some 5,800 xerafins per year. But he met no matching response, the religious in particular claiming their allowances sacrosanct and refusing to give up anything. Linhares did later proceed with the reduction of some salaries and the abolition of a few redundant minor offices, but with little effect on the overall financial state of the viceroyalty.[20]

Even less fruitful were his attempts to persuade the crown to allow him to consolidate all government revenue at Goa into a central fund to be drawn on for expenses of state in order of strict priority, military and other urgent and essential requirements taking precedence over all other claims. Because the crown had committed certain specific revenues to the church in Asia which, under the terms of agreement with the papacy, it was bound to support, and because it regarded the conversion and spiritual care of Asians as its principal responsibility in the region, Linhares' request for a consolidated fund was rejected.[21] Thus, like his predecessors and his successors in the Hapsburg years, this viceroy found himself in the position of having to give priority to the payment of salaries and allowances to an unnecessarily large ecclesiastical establishment at a time when the Estado da Índia was struggling, on demonstrably inadequate resources, for its very survival.

Much more determination was shown by the home authorities in attempts to eradicate slothfulness and corruption among its officers in Asia, but again with little success. It was often claimed by contemporaries that the financial distress of Portuguese Asia was largely the fault of lazy and corrupt officials, and that in fact the revenues locally available would have been quite adequate if administered properly. At the beginning of the seventeenth century the viceroy Count of Vidigueira had told the crown that expenses "would easily be accommodated" if all income from the various parts of the State of India, apart from that required locally for the upkeep of forts and factories, were forwarded to Goa.[22] Francisco Pais likewise claimed that income should have exceeded expenditure in the State of India, but that the reverse was the case because of "greedy irregularities committed by officials with little zeal for His Majesty's service."[23] He went on to argue that local income ought to have been quite sufficient to finance wars and extraordinary emergencies as well as routine matters, citing the records of several past viceroys and governors from Dom Constantino de Bragança (1558-1561) to André Furtado de Mendonça (1609) as examples of what could be achieved. Another critic of the period asserted that the crown was robbed of over 200,000 cruzados in duties every year, while in 1629 the crown itself ordered Viceroy Linhares to institute a special inquiry into state finances on the grounds that the revenues of the Goa territories ought to be showing a surplus of over 57,000 xerafins.[24]

There is no doubt that much corruption did in fact exist in the financial administration of the State of India and that it was particularly acute in regard to the payment of duties. Pyrard observed that export duties on goods despatched from Goa to Lisbon would be assessed at "very little" if the necessary deal were arranged with the *vedor da fazenda geral.* Upon this latter official, according to Viceroy Linhares, "the whole Treasury of Your Majesty in India is entirely dependent," and he could consequently "play the thief" more effectively than any other bureaucrat in the Indies—apart from the viceroys themselves.[25] Other alleged malpractices of the early seventeenth century included the illegal sales of pepper, cinnamon and similar high value commodities by Portuguese officials and their accomplices at such places as Ormuz, Muscat and Surat, a practice viceroys were personally accused of aiding and abetting.[26] Tax-contractors predictably figured in the financial scandals at Goa. Bartolomeu Sanches Correia, who farmed the Goa customs from 1629 through the early 1630's, was accused of embezzling large sums from both the crown and the city council, and of paying too little for the original contract.[27] Slackness in the accounts department (*casa dos contos*) owing to the employment of allegedly corrupt and incompetent favorites of the viceroys, overpayment and excessive numbers of officeholders, evasion of registration regulations for goods shipped to Europe, allowing excessive time to the state's debtors to settle their debts while meeting fraudulent claims by its so-called creditors—all helped to compound further the financial distress of the Goa government.

Linhares alleged that the clergy in Goa "assume that anything belonging to the King can be stolen and concealed with a very good conscience."[28] He also claimed that even when responsible higher officials were honest, widespread peculation was inevitable since "all government workers are thieves" and the higher officials could not personally supervise everything.[29] Linhares' diaries are full of comments about the laziness, dishonesty and incompetence of officials in India, and the mediocrity of his viceregal councillors. The situation at the Portuguese fortress in Chaul provides a typical example. Although the crown was paying for sixty soldiers to man this fortress, it was revealed in 1631 that the garrison in fact consisted of eleven men only. The building itself was "dirty and broken-down," its field of fire obstructed by houses and palm-groves, its magazines without gunpowder or munitions, and its despatch boat in an unseaworthy condition. The fort captain was never present, no proper guard was kept, the keys were in the hands of private individuals and anyone could go in and out at will. "And if I now should wish to punish these things," Linhares commented bitterly, "I immediately gain more enemies—and if I do not punish them the fortresses are put at risk."[30] Similar strictures regarding the unpreparedness of fortresses in the State of India abounded in the instructions and reports of other viceroys and governors, especially stressing the lack of artillery and gunpowder, inadequate and inefficient garrisons, and the prevailing disrepair of walls and defense works.[31]

The extent of corruption in seventeenth-century Portuguese Asia obviously cannot be accurately measured, but the frequency of complaints suggests it was deep and widespread. In his study of "the structural crisis of European-Asian trade in the early seventeenth century" Niels Steensgaard argues that the Portuguese were concerned less with monopolizing commerce than with controlling and preying upon it, that their officials were remunerated more by a share of the prey than by salaries, and that to eliminate corruption was therefore impossible without paradoxically eliminating the Estado da Índia itself.[32] This view is consistent with the indifference or outright hostility to reform shown by many officeholders which, with the lack of means to enforce changes and the non-availability of competent replacements for those found guilty of misconduct, made many abuses virtually ineradicable even by the most zealous of viceroys.[33] Despite these chronic weaknesses, however, it is unlikely that Portuguese officials of the seventeenth century were substantially more addicted to fraud than their predecessors, and corruption per se should not be regarded as the decisive causative factor for the financial crisis in the later Hapsburg decades. The fact was that these decades saw a relative decline in the State of India's revenues and a simultaneous increase in its expenditures, which would have occurred whether corruption had been rife or not.

As well as considering how the revenues available to government in the Estado da Índia could be made to go further by economizing, rationalizing and combating graft, the Portuguese also attempted to provide additional funds through new taxes, the raising of loans and the provision of special aids from the home government. From 1593 a special levy called the *consulado* had been imposed on goods entering and leaving Goa and other Portuguese Asian cities, and from 1623 a *collecta,* or tax on rice and other provisions, was introduced at Goa on the initiative of the leading citizenry.[34] Both imposts were intended to raise funds for city defenses, but the yield was apparently inadequate to fulfill this purpose. Many cities failed to maintain their walls and fortifications, Goa herself remaining poorly defended, her revenues insufficient even for such items as paying the salaries of municipal officials, caring for lepers and orphans, and providing food at a reasonable price for the poor.[35] Thus the consulado and collecta were only palliatives which partially relieved the viceregal authority of one of its defense burdens, but did not solve its fundamental difficulties.

The raising of loans, either from corporations such as the *misericórdias* and the câmaras, or from individuals, was in practice the most substantial means of obtaining additional ready cash for the viceregal government. By the end of the first decade of the seventeenth century the royal treasury at Goa apparently owed 600,000 cruzados, a sum which undoubtedly increased as the years passed.[36] However, the Goa government had a low credit rating, so that Viceroy Linhares repeatedly found himself unable to raise urgently needed cash except on his own personal surety "because what is loaned for the King's service is never repaid."[37] Force, or at least strong pressure tactics, often had to be used to extract loans from reluctant creditors. Nevertheless loans were raised. The misericórdia, which at its various establishments in Goa, Macao, Malacca and elsewhere maintained considerable cash funds, largely as a consequence of its trust and banking activities, was probably the chief source. In 1623 the crown itself instructed Viceroy Vidigueira to borrow the comparatively large sum of 200,000 cruzados from the trustee fund of the Goa misericórdia, to be repaid at Lisbon in the form of bonds.[38] The heirs to the estates thus seized were presumably not consulted. In 1630 the naval commander Nuno Álvares Botelho took money on his own responsibility from the misericórdia at Malacca to pay and provision his men. The viceroy considered this move acceptable, but to quieten complaints he arranged for the treasurer to present a written acknowledgment of the debt to the misericórdia at Goa. Viceroy Linhares also borrowed at various times from the câmaras of Goa and Bassein and sought what amounted to a subsidy from the câmara of Macao. According to Viceroy Aveiras, money from the Goa misericórdia and other sources had been exhausted, even before he arrived in the city in 1640.[39]

Individuals were wrung for money in a variety of ways, usually by being required to fit out vessels and supply troops for expeditions, or simply to provide hard cash. The Portuguese and Hindu communities at Goa both loaned funds to finance Viceroy Martim Afonso de Castro's fleet to relieve Malacca in 1605, and 5,000 xerafins were still outstanding to the Hindus eight years later. In order to raise a force for the recovery of Mombasa in 1631 Viceroy Linhares extracted, under threat, a series of loans mostly of between 1,500 and 3,000 xerafins each, from wealthy persons at Goa, at least one of whom subsequently complained to Lisbon over non-repayment. Viceroys and governors themselves were usually the greatest contributors to such enterprises. They raised loans on their own credit although often, as in the case of Viceroy Rui Lourenço de Távora (1609-1612), in the expectation that the state would eventually assume responsibility. In 1617 Viceroy Dom Jerónimo de Azevedo pawned his gold and silver to provide money for the annual pepper fleet to Cochin, while Viceroy Linhares personally contributed 22,500 xerafins for the 1631 Mombasa expedition, raising this sum through the sale of his wife's proprietary rights to the East African fortress of Sofala. Like Linhares, Viceroy Aveiras claimed in 1640 that he sustained his government only on his own credit.[40]

It was of course not unusual in the seventeenth century for viceroys and governors to spend parts of their personal fortunes on state affairs, just as it was common practice for them to use the opportunities of office to reimburse themselves. However, reliance on the viceregal pocket in an imperial possession so short of ready cash as the Estado da Índia in later Hapsburg times could be carried to excess—or so it seemed to Linhares, who wrote to Lisbon, in a moment of frustration perhaps, that having exhausted his personal fortune on behalf of the state he had been reduced to the personal embarrassment of wearing torn shirts, and there was now not a single man in India who would lend him 100 cruzados. Linhares, who later returned to Europe a wealthy man, was exaggerating, but the blunt demand with which he ended this testy letter, "that I be given precise instructions what I should do," required an answer which was never satisfactorily given.[41]

Clearly, it was impossible to provide the Estado da Índia with all the funds it needed over and above revenue from locally raised loans. Therefore the crown was obliged to resort to an expedient which, from its own viewpoint, must have seemed particularly inappropriate and distasteful. This was the provision of regular cash subsidies, usually in the form of silver or gold, shipped out on the annual carracks from Portugal. The money was pre-assigned to particular defense needs and, from at least the first decade of the seventeenth century, had become a substantial and essential part of the Goa government's income. The need for such aids bears out the claim that the cost of the State of India to the Portuguese crown was by then greater than its value. The subsidies known to have been despatched from Lisbon to Goa in certain years of the late Hapsburg period (Table 5) were not large in relation to the total needs of the Estado da Índia, but as hard cash-in-hand reserved exclusively for military purposes they were vitally important. They were also certainly difficult for an impoverished and overcommitted government in Lisbon to find, and the difficulty was heightened and made more ironical by the marked decrease in returns at Lisbon itself from the intercontinental India-Portugal trade during the first few decades of the seventeenth century. The quantity of pepper exported annually through Goa to Portugal dropped sharply in this period, and by the late 1620's and early 1630's was averaging less than half the sixteenth-century level. A simultaneous decline in the selling prices of pepper at Lisbon from the 1620's,

Table 5. Cash Subsidies Despatched from Lisbon to Goa in 1626 and 1630-1635 (in cruzados)

Year	Amount
1626	80,000
1630	52,000
1631	80,000
1632	80,000
1633	170,000
1634	50,000
1635	80,000

Sources: BN Rio, codex 2/2-19, f.59; AIG, codex 2358, ff.303 and 307, and codex 2608, ff.43, 47v, 49, 50v.

and a fluctuating but predominantly rising trend in buying prices at source in Malabar and Kanara in the 1620's and 1630's, accentuated the decrease in returns.

Meanwhile, the risks of the Goa-Lisbon trade were much greater in the Hapsburg years—partly owing to the Dutch challenge and the depredations of Algerine corsairs, but more especially because of deteriorating conditions on the carreira da Índia, with more abortive voyages and an increasing rate of shipwreck on the home run from the Mandovi to the Tagus. As a result, Lisbon and European entrepreneurs became increasingly cautious about participating in Portugal's Asiatic trade, from the end of the sixteenth century. As early as 1597 the Portuguese crown found it impossible to interest anyone in accepting its proffered five-year contracts to run the pepper monopoly, and the same unwillingness to become involved in the official Portuguese Indies trade was prevalent when plans for a Portuguese India Company were under discussion in the early 1620's.[42]

Finally, the obvious if short-sighted policy of selling offices on a massive scale also provided no real solution to the State of India's financial difficulties. A drastic attempt to raise defense funds through a general auction, the purchasers to be allowed to take up office before those already on the waiting lists, was ordered by the crown in 1615. This expedient was strongly resented by fidalgos and soldiers serving in Portuguese Asia, many of whose claims to office were retarded in favor of the purchasers, but it was apparently repeated in 1637-1638.[43]

Dom Cristovão de Moura, the viceroy of Portugal between 1600 and 1603, once remarked in a letter to the king at Valladolid that the three things essential to defend the Estado da Índia were "money, men and ships."[44] If a shortage of cash had been the only serious problem facing Portuguese governments in Asia in the later Hapsburg years, then a slim chance that the crisis would ultimately be weathered, as it was in northeastern Brazil between the 1620's and 1650's, might have existed. However, with the persistence of military pressure from steadily strengthening Dutch and English forces, it was Portugal's inability to produce the shipping and manpower resources needed to defend her Asian interest, in combination with the economic crisis, that made the imperial losses of the mid-seventeenth century inevitable.

From the first appearance of European rivals in Asian waters, the Portuguese had predictably adopted an attitude of implacable opposition. In March 1595, without even the usual prior clearance with the king in Spain, the governors in Lisbon despatched an urgent letter to Goa, warning of the preparations for Cornelis Houtman's voyage underway in Holland and urging the viceroy to ensure that the Dutch, if they reached their destination, never returned home to Europe, but instead received "the punishment which they deserve."[45] In fact Houtman's voyage was successfully accomplished and, together with those of his successors over the next few years, plainly demonstrated how much easier it was for the Portuguese to formulate an uncompromising attitude against the European newcomers than to enforce it. Nevertheless, the Lisbon government made strenuous efforts to provide the Estado da Índia with the ships and men it needed. In 1604 five large carracks and galleons under a new viceroy, Dom Martim Afonso de Castro, were despatched to Goa, and in 1605 ten more under Braz Teles de Meneses and Álvaro de Carvalho. An even larger squadron prepared in 1606 was unable to sail because of a Dutch blockade of Lisbon, but in 1607 seven ocean-going warships left for Portuguese Asia. Finally, in 1608 one of the largest fleets that had ever left the Tagus for India, composed of fourteen carracks and galleons under the overall command of the Count of Feira, viceroy-designate, was despatched in a major effort to swing the balance against the Dutch, greatly straining Portugal's limited resources in the process. It was followed by five more sail later the same year.

Partly through bad luck, partly through mismanagement, but primarily because of the good leadership and ample resources at the disposal of the Dutch, these Portuguese efforts failed in their primary purpose. Over one-third of the Portuguese ships sent to India in the five-year period 1604-1608 were wrecked, captured or burned by the Dutch in preliminary encounters near Mozambique, or forced to turn back home. A series of battles fought off Malacca in 1606 and 1607 was indecisive, while the fleet of the Count of Feira lost its commander when still in the Atlantic, was poorly handled and proved unable to accomplish any important successes against the enemy. The mounting alarm in Lisbon and Valladolid at this turn of events was well justified and was not allayed, particularly in the Portuguese capital, by the truce which came into effect in Europe in 1609.[46]

The Twelve Years' Truce, which was secretly arranged by Philip III's Spanish minister, the Duke of Lerma, and of which the Portuguese were told little before it was concluded, met strong opposition from both the Council of Portugal and the viceroy in Lisbon, Dom Cristóvão de Moura. As the Portuguese foresaw, it did not stop the Dutch activities in Asia but rather strengthened the Netherlands East India Company's ability to sell its spices in Europe where it now received access to Iberian markets. At the same time the crown had every intention of continuing the struggle beyond the line and from the first instructed the viceroy at Goa to continue efforts to expel the Dutch as long as their presence in Asia persisted.[47]

Hitherto the English—with more modest resources at their disposal in Asian waters at this time than the Dutch, despite their slightly earlier commencement of direct trade in the region—had sought to avoid open conflict with the Portuguese.[48] However, fighting between the two nations took place off Surat in 1611, and off the same town in 1614 an English squadron under Sir Thomas Best inflicted severe losses on a Portuguese force of galleons sent against it from Goa. The struggle reached a climax ten years later when English naval support enabled Shah Abbas to capture from the Portuguese the important strategic and trading port of Ormuz on the Persian Gulf. This was the first major territorial loss the Portuguese had suffered in their wars against the northern European newcomers, and marked a dramatic and ominous beginning to the critical 1620's and 1630's.[49]

With the end of the Twelve Years' Truce (1621) and the fall of Ormuz in 1622, the inability of the governments in Goa and Lisbon to provide the Estado da Índia with the naval forces it so desperately required, had become painfully obvious to informed observers. A realistic report to the king a few years later stated frankly that by 1622 Portugal's enemies beyond the Cape of Good Hope had become "absolute masters of the sea," their eighty ocean-going carracks and galleons overwhelmingly outnumbering Portugal's seven or eight. Reminding Madrid that the State of India had previously been able to field some 15-20 galleons, 10-12 war galleys and 150-200 smaller warships, the author of this report continued that time and war had so depleted these forces that only eight galleons, two galleys and sixty foists remained. To restore the situation the crown would have to maintain thirty well-armed and properly manned galleons in Indian waters, and a further twenty based on Malacca, the latter to be reinforced, if necessary, by Spanish forces from the Philippines.[50]

These demands were beyond the ability of the Portuguese government to meet in the early 1620's, and even more hopelessly so after the Dutch invasions of Bahia (1624) and Pernambuco (1630) which forced Lisbon to concentrate all available naval forces off Brazil. In the event, between the arrival of Captain-Major Nuno Álvares Botelho's five-galleon squadron at Goa in 1624 and the end of Hapsburg rule in Portugal in 1640, only one substantial reinforcement of ocean-going warships reached the Indian Ocean from Lisbon, and that was four of six galleons that the Count of Linhares took with him on proceeding to Goa as the new viceroy in 1629.[51] Nor was it realistic to expect the State of India itself to contribute more than one or two galleons every few years from Indian dockyards.[52] It is understandable, therefore, that Linhares should have demanded plaintively of the crown less than four years after his arrival, "what can six galleons and two pinnaces do against twenty, thirty or more enemy carracks?" The viceroy went on to request that he be sent twenty-four well-equipped galleons, warning that if these were not provided immediately, a truce with Portugal's European enemies in Asia would have to be made.[53] The galleons did not come, and within two years Linhares had signed his truce with the president of the English East India Company at Surat. This at least neutralized the English, but it did not remove the still more lethal threat from the Dutch. In the final year of Hapsburg rule in Portugal the new viceroy at Goa, the Count of Aveiras, found only two serviceable galleons available in the viceroyalty, yet vainly requested a minimum squadron of ten to twelve such warships if the struggle against the Dutch were to be continued. Since an effective cessation of hostilities with the Dutch East India Company was not achieved until 1663, it was hardly surprising that the Dutch should have deprived the Portuguese of most of their possessions in southwest India and southeast Asia in the intervening years.[54]

Portugal's chronic lack of warships in Asian seas in the Hapsburg years, and of ready cash in the viceregal treasury, was matched by her shortage of manpower. "If our need for money is great," wrote Viceroy Linhares to the Count of Castro d'Aire in Lisbon in 1632, "then the need we suffer for soldiers, sailors and gunners is even greater."[55] The shortage was also felt, and particularly acutely, at the leadership level, there being too few Portuguese fidalgos of proven ability available to fill high military positions, in the absence of suitable grants of office and other privileges to attract such men to an Estado da Índia straitened in circumstances.[56] The extent of the manpower shortage in the critical late 1620's and 1630's can be gauged with some precision, if Viceroy Linhares' own figures may be credited. On the basis of a fortress-by-fortress and squadron-by-squadron survey, he informed the crown in late 1634, after five years' bitter experience in office, that at least 7,192 soldiers were required to man the defenses of the Estado da Índia, excluding troops needed for emergency expeditions. In fact, there were less than 3,000 Portuguese soldiers in the whole region.[57] Repeatedly during his term as viceroy Linhares urged that, to garrison the fortresses and armadas alone, it was absolutely essential to despatch immediately a reinforcement of 4,000 men from Lisbon, and that this would have to be regularly backed up by a further 1,000 every year.[58] These requests, like the viceroy's demands for more galleons, were never adequately met. Although no other viceroy in the mid and late Hapsburg years spelled out the problem quite so explicitly as Linhares, most of them complained repeatedly of the paralyzing effect of the manpower shortage. The Count of Redondo (1617-1619), the Count of Vidigueira (1622-1628), Pero da Silva (1635-1639) and the Count of Aveiras (1640-1645) stressed particularly the lack of gunners, seasoned soldiers and skilled seamen, but received little satisfaction from Lisbon.[59]

For a variety of reasons the Portuguese forces in Asia suffered very high rates of death, disablement and desertion in the Hapsburg years. However, the fundamental causes for the manpower shortage lay less in conditions in the Asian possessions themselves than in the inability of the home authorities to raise recruits in Portugal, and their failure to transport safely to their destination such contingents as were raised. In the later Hapsburg era Portuguese Asia had to compete for its European recruits in a severely limited market, not only with Brazil, then under partial occupation by the Dutch West India Company, but with Philip IV's regiments engaged in the insatiable wars of Flanders and Italy. Thus the Portuguese authorities seldom succeeded in assembling the minimum number of soldiers needed at Goa in any particular year, despite strong urgings from the crown. Even such expedients as calling on the magistrates in districts round Lisbon to raise levies, sending recruiting agents to Braga in the northernmost province of Entre Minho e Douro, attempting to force men of the marine regiment to transfer to the Goa service, and sweeping the Limoeiro prison, failed to yield the manpower required.[60] Moreover, there was frequently a marked disparity between the numbers of soldiers officially enlisted in Lisbon and those subsequently recorded as having arrived at Goa. Apart from the possibility of fraudulent rolls being drawn up at the Casa da Índia before departure, the main reasons for this were the high rate of mortality often experienced on the long voyage round the Cape and the inclusion of varying proportions of militarily useless pre-adolescent boys instead of men among the contingents shipped.

Filippo Sassetti, the Italian factor representing Giovanni Battista Rovellasco and his fellow pepper contractors at Cochin in 1583-1586, had asserted that of 2,500-3,000 men and boys who had left Portugal for Asia per year in this period, from a quarter to a half died en route.[61] Sassetti may perhaps have exaggerated somewhat, but half a century later, of the unusually large number of 1,431 soldiers officially embarked on three carracks for India in 1633, 171 were listed as having died before arrival at Goa, or almost one-eighth of the total. Moreover, many of the survivors from this voyage arrived ill—and a high mortality rate tended to persist in the hospitals of the viceregal capital and in the remoter outposts beyond.[62]

Almost as worrying a problem for the recruit-hungry Goa authorities was the illegal but well-entrenched custom of shipping out large numbers of boys, many aged six or seven years or even less, and including them in the official military lists. The boys were sent to India either as recruits for the religious orders or as pages, but they often suffered severely on the long and uncomfortable voyage from Europe and many subsequently died in hospital at Goa. What proportion of "soldiers" embarked in Lisbon was likely to turn out at Goa to be actually pre-adolescent boys on an average shipment is uncertain. However, two examples should indicate the seriousness of a practice which might have been allowed to pass in earlier years when the manpower shortage was less acute, but could no longer be tolerated in the 1630's.[63] In May 1632 the captain-major of the India carracks, António de Saldanha, complained that he had found a large number of small boys on his flagship, and that the other two carracks in the squadron also included many. A team of investigators had to be sent to inspect the ships and disembark such boys as could be found.[64] On the ship in which Pero da Silva sailed to India in 1635, thirty small boys were put ashore before the vessel set out—but over ninety more were still found to be on board after the voyage had begun, while on the accompanying carrack there were even more. Pero da Silva's complaints on this particular occasion caused the crown to reiterate in 1638 that no child under the age of thirteen should go to India.[65]

Whatever their causes, the shortcomings of the Portuguese system of recruitment and transportation of soldiers for India in the late 1620's and 1630's are well brought out by comparing the enlistment figures for Lisbon and Goa respectively. Between 1629 and 1634, when the Count of Linhares was viceroy at Goa, 5,228 soldiers had been embarked in Portugal for Asia, according to the official rolls drawn up in Lisbon—yet the Goa records showed that only 2,495 had actually arrived. This was a reduction of 52 per cent. Moreover, in 1633, which had been the best year for recruits in this half-decade, only 547 newly arrived men from Portugal actually enlisted for active service in one or another of the various fleets despatched to Malacca, Muscat, the Kanara coast and elsewhere—out of the alleged total of 1,431 "soldiers" embarked at Lisbon.[66] These figures help to explain why Viceroy Linhares asked for such large reinforcements in relation to the numbers actually required for the India service.

The effects of the manpower shortage on Portuguese military activities in Asia in this period often proved crippling. When Dom Felipe de Mascarenhas, who had been sent to replace Constantino de Sá as Portuguese commander in Sri Lanka in 1630, asked the viceroy for 1,000 Portuguese soldiers, Linhares responded that since he could not muster even fifty men at Goa for any enterprise, he would try to arrange for a few hundred to be assembled together from the garrisons in Cochin and the Kanara forts.[67] Three months later Linhares twice had to scour his own expeditionary fleet to the Malabar coast, as well as neighboring forts, and to offer the inducement of payment in gold, to obtain just 160 reinforcements for Sri Lanka. On another occasion he offered what he claimed were the highest wages ever given to soldiers serving in India to attract men for service in Sri Lanka, but two days later had to fall back on recruitment by force.[68]

The limitations imposed by the manpower shortage are also underlined by Linhares' failure to establish and maintain a *terço* at Goa, although the crown had specifically advised him that this was one of the principal tasks for which he had been appointed.[69] The creation in Goa of a terço organized on the Castilian model had been long recommended and several times unsuccessfully attempted. Linhares' terço was planned on paper as a unit of 2,500 men to be made up of twenty companies of 125 soldiers each, but it never approached this total. The few companies actually formed were repeatedly depleted to provide reinforcements for Sri Lanka, Malacca and elsewhere. In 1634 Linhares wrote informing the crown that the terço no longer existed since the treasury had no money to maintain it, and added that in so far as it had functioned at all in previous years it had done so at his own personal cost.[70] The following year the new viceroy, Pero da Silva, told the crown that it was impossible to maintain the terço as there were too few men available, and because "I cannot see from where the wages of its soldiers can come."[71] The experiment was not successfully revived at Goa until 1671. Ultimately the only way to alleviate the chronic shortage of military manpower in Portuguese Asia was to recruit local Asians and Africans, and in later Hapsburg years this was done to a considerable extent. When Linhares finally succeeded in putting together a respectable force of almost 2,000 expeditionaries for the Sri Lanka campaign in October 1631 hardly any of the men were Portuguese, over 400 being Negroes and most of the rest Lascars and Canarins, despite the low martial reputation of the latter.[72]

Portuguese Asia was clearly facing critical difficulties in the later Hapsburg period—but the home government, although well aware of these difficulties, refused to accept them as irremediable. In the early 1620's it began to examine seriously the proposition that the Goa-Lisbon trade could be revived—and the difficulties of the State of India correspondingly ameliorated—by the creation of a monopoly trading company. The English and Dutch had sustained their unwelcome commercial intrusion into Asian seas by forming just such companies, which were widely acknowledged to be among the most modern and successful trade organizations of the era, capable of utilizing capital and expertise in a highly effective manner. If Englishmen and Dutchmen, who were both greenhorns in the difficult business of direct trade with the Orient, could operate these enterprises successfully in that region, there seemed good reason to believe that the Portuguese, experienced veterans of the East, could do likewise—and perhaps beat the interlopers at their own game. At the least, if Portugal could modernize her trading organization for Asia, she could compete more effectively with the newcomers on the commercial level. The formation of a chartered joint-stock company in 1628 to assume control of the official Portuguese India trade was a serious attempt to achieve precisely this objective.

5 / A Company Conceived

On the return of Vasco da Gama from his pioneer voyage to Calicut in 1499 the Portuguese government had formed a syndicate for the exploitation of the Cape trade route, in which both the crown itself and certain private interests participated. Although this syndicate apparently did not exercise a monopoly of the trade, it was probably responsible for organizing the annual voyages to and from India for the five-year period 1499-1504.[1] Subsequently a more discriminatory policy was introduced, aimed at maximizing profits for the crown. From 1506 the provision of Indiamen, the trade in precious metals from Portugal to India, and the trades in pepper and other major spices in the reverse direction, were reserved as royal monopolies.[2] Actual practice was apparently less restrictive than these regulations suggest. Private as well as crown shipping was used on the Cape trade route, and a substantial private trade in non-monopoly goods such as Asian textiles and gems came into being. Legitimate private traffic in monopoly products was carried on by naval personnel, and by certain privileged institutions and individuals under royal license, and there was also an indeterminable volume of smuggling.[3] With these exceptions and qualifications the monopoly system could be said to have continued in principle for over half a century until the 1560's when the crown, confronted with mounting liquidity problems, began promoting ways of increasing participation by private enterprise.

In pursuance of this aim the pepper monopoly itself was briefly contracted out to a merchant syndicate in 1564. Then in 1570 the crown issued a new set of standing orders which opened the trades in pepper and other spices to free competition, although it continued to trade in these commodities itself and to retain its monopoly on the export of precious metals to Asia. This experiment was short-lived and was soon superseded by a more thorough-going system of contracts. Between 1576 and 1597 the pepper monopoly was farmed out to a succession of syndicates composed variously of German, Italian and Portuguese interests, usually for five-year periods. In the last quarter of the sixteenth century the monopoly trades in other spices, the provision and outfitting of the India fleets, and the collection of freight and customs charges at Lisbon were also farmed out, usually separately but sometimes in combination to the same contractors.[4]

Although the contract system did provide a temporary relief for the cash shortage in the Portuguese Asia trade, it proved unsuccessful as a long-term solution. The pepper syndicates in Lisbon, largely for reasons beyond their control, were consistently unable to import the quantities specified in the contracts and generally failed to make an adequate profit. Two of the most prominent financiers involved, the German Konrad Rott and the Italian Francisco Rovellasco, were forced into bankruptcy. From about the end of the sixteenth century, especially following the English and Dutch intervention in the seaborne spice trade, private enterprise became unwilling to take up further Portuguese pepper contracts, although the more profitable farms on freight and customs continued to find takers until 1616. Thus although its cash shortage was becoming more acute than ever, the crown was reluctantly forced to resume its trade monopolies and, at least in part, its shipping monopoly.[5]

Under these circumstances efforts naturally continued in the early seventeenth century to find acceptable ways of involving private capital in the official India trade. Following the commencement of the Twelve Years' Truce with the Dutch in 1609, the suggestion was made to open the Lisbon-Goa trade to merchants of all nations on condition that returns were divided equally with the crown. A proposal was also considered to offer free trade and reduced tariffs to all Portuguese. However, neither of these ideas was apparently thought practicable enough to warrant trial.[6] Instead, over the following decade the crown began to show increasing interest in the possibility of sponsoring a joint-stock company, patterned on the models of the English and Dutch East India Companies.

It is impossible to say for certain when the formation of a joint-stock company to operate Portugal's Asia trade was first proposed, or by whom. Some historians argue that a short-lived India company was actually formed by Philip II as early as 1587, but this claim carries little conviction.[7] The probability is that such an enterprise received serious consideration in Lisbon and Spain only after the English and Dutch companies had proved their viability in the early seventeenth century.

In 1619 the English adventurer Sir Anthony Shirley, then Shah Abbas of Persia's ambassador to Spain, was suggesting the establishment of an India company "in imitation of that of Holland." Similar proposals made by the New Christian merchant Duarte Gomes Solis, though not finally published until 1622, had probably been circulating in manuscript for the previous ten years. By about 1618 the then viceroy of Portugal, Dom Diogo da Silva de Mendonça, Duke of Francavila and Marquis of Alenquer, was lending his support to the proposal.[8] Francavila's advocacy may have finally persuaded the Iberian crown to try the scheme, but in any event on February 19, 1619, the crown made it known that an India company was to be formed. Investment in the company was to be widely encouraged and the Portuguese câmaras were asked to set an example by publicly lending their support.[9]

No further steps appear to have been taken towards establishing a company for almost five years. The delay can probably be explained by the sudden changes of political climate in Madrid and Lisbon which followed the death of Philip III on March 21, 1621, ending the political ascendency of the Duke of Uceda. The new monarch, Philip IV, a youth of sixteen, soon began to entrust affairs of state to his former tutor Gaspar de Guzmán, then Count of Olivares, and from 1625 Count-Duke. Within a year Olivares had established himself as undisputed chief-minister and had become the effective arbiter of all major decisions of state, including those which concerned Portugal. The Duke of Francavila was dismissed as viceroy at Lisbon and replaced by a board of governors composed of Dom Diogo de Castro, Count of Basto, and Dom Diogo da Silva, Count of Portalegre, who later opposed the formation of an India company.[10]

At the same time the resumption of full-scale hostilities with the Dutch on termination of the Twelve Years' Truce in 1621, and the fall of Ormuz in 1622 to Anglo-Persian forces, gave new urgency to the provision of financial aid for the defense of Portuguese Asia. This exigency tended to divert attention and money from the company. The Lisbon Câmara, which only in 1619 had been obliged to find an extraordinary 200,000 cruzados to help finance a leisurely tour of Portugal by Philip III, was now asked for a further 200,000 cruzados for the war effort in the Persian Gulf, money which might otherwise have been available for investment in the company.[11] Moreover, a temporary sharp reduction in silver imports from America to Seville in 1622-23 probably lessened the chances of government funding during that period.[12]

Nevertheless, in 1624 the plan for establishing a Portuguese India company was revived—for reasons not entirely clear, but seemingly connected with the return to court that year of Dom Jorge Mascarenhas. Dom Jorge—who appears to have been first converted to the scheme by Father Belchior de Seixas, a Jesuit living at Goa[13]—quickly became its leading spokesman, and in 1624 the Council of Portugal recommended that he be appointed chairman of the company in the event of its formation. Thereafter, despite considerable criticism, he remained strongly committed to the company, probably partly as a means of promoting his own influence at court against that of rivals such as the Duke of Villahermosa and the Counts of Basto and Portalegre.[14]

Lobbyists for the India Company were greatly assisted by a simultaneous and important new development in Iberian foreign policy. In 1624 Olivares was in the process of evolving a "Great Project" designed primarily to disrupt and if possible to destroy Dutch trade in the Baltic, in the same manner as the Dutch themselves had been harassing with such effect Spanish and Portuguese trade with America and Asia.[15] The linchpin of this scheme was to be the formation of a Hapsburg-Hanseatic trading company financed largely by German and Flemish merchants in cooperation with the Iberian and Imperial crowns. This company would receive a monopoly of trade between the Baltic and the Iberian peninsula to the particular exclusion of the Dutch. It would also be involved in certain military operations, especially the reoccupation of ports in East Frisia that had been seized by the Dutch in 1623. More broadly, the "Great Project" envisaged a whole series of interlocking companies to serve northern Europe, the Middle East, Spanish America and Portuguese Asia, thereby not only weakening the Dutch but reinvigorating Iberian international trade in general.[16] The forming of a Portuguese India company to compete more effectively with the Dutch and English in the Asia trade thus became desirable to Madrid, as part of a worldwide strategy. The first important step towards launching it was then taken on July 13, 1624, with the appointment of Dom Jorge Mascarenhas as president of the Lisbon Câmara.[17] The declared purpose was to rally câmara support for the company.

The career of Dom Jorge Mascarenhas within the royal bureaucracy in later Hapsburg Portugal was remarkably varied and eventful, and merits description in some detail. Although he held some of the highest offices of state under three kings and would be elevated to a marquisate, Dom Jorge came from the lower ranks of the nobility. Born probably in the 1580's he was descended on his father's side from the lords of the village of Mascarenhas in Trás-os-Montes. His father, Dom Francisco Mascarenhas, had been captain of Ormuz, and there was a strong family tradition of service in the Estado da Índia.[18] Like most contemporary fidalgos Dom Jorge began his public service as a military officer. He was colonel of a regiment by 1602, when he was also appointed keeper of the royal household to Philip II. Four years later he had given sufficiently distinguished naval service to receive a letter of thanks from the king. In 1615 he obtained his first important command, the captaincy of Mazagan in North Africa, some time after having been made a member of the council of state.[19] While returning from Mazagan in 1619, his ship was attacked by Barbary corsairs and, despite desperate resistance, he was taken captive. He was ransomed from Algiers by Philip III, and then appointed governor of Tangier in 1622, another frontier posting. By 1624 he was back in Portugal where he was created Count of Castelo Novo and made governor of the Algarve, the same year as his appointment to the presidency of the Lisbon Câmara.

Thereafter Dom Jorge received a series of important offices concerned with the administration of Portugal's overseas territories. He became chairman of the India Company's board of directors in 1628 and was also made president of a treasury board which was charged with regulating the Brazil trade, primarily for the purpose of raising money to fight the Dutch in Pernambuco. The treasury board lasted only two years and was disbanded along with the Portuguese India Company itself in April 1633.[20] These failures apparently did not greatly affect Dom Jorge's standing at court, for he continued to receive the confidence and favor of the crown and was made first viceroy of Brazil in 1639, and Marquis of Montalvão in 1640. He retired from Brazil two years later, having supported the restored Bragança monarchy and ensured the adherence of Portuguese America to Lisbon rather than Madrid. Back in Portugal, he was soon appointed president of the Overseas Council (*Conselho Ultramarino*), in effect the crown's chief advisor on the colonies. However, because his wife and sons had remained loyal to the Hapsburgs Dom Jorge was not fully trusted by the Bragança regime. He was several times arrested and eventually died a prisoner in the castle of São Jorge in Lisbon in 1652.[21]

In 1624 an ad hoc committee chaired by Dom Jorge Mascarenhas was appointed to work out details, in accordance with general guidelines drawn up by the Council of Portugal, for formation and organization of the India Company.[22] Other members of this committee were Dom Jorge de Almeida, later admiral of the India Fleet and captain-general of Sri Lanka; João de Frias Salazar, a high court judge and councillor in the Lisbon Câmara; Diogo das Póvoas, the director of customs; and Leonardo Fróis. Three or four merchants with suitable experience were also to be co-opted by the governors of Portugal on Dom Jorge Mascarenhas' advice. The committee was to formulate recommendations which would in the first instance be passed on to the governors for comment and then sent to the crown.

The most urgent and difficult task facing the committee was the raising of capital. The Lisbon Câmara, as the country's most important municipal council, was asked to give a lead by declaring without delay an amount it was prepared to subscribe. Probably as a result of pressure from Dom Jorge Mascarenhas, who had deliberately been given the double role of chairman of the ad hoc committee and president of the câmara, the response was prompt and reasonably positive. By January 1625 the câmara had agreed in principle to give support, and within a month was taking steps to raise the necessary funds through a new issue of municipal bonds.[23] Delays and difficulties mostly not of the câmara's making interrupted implementation of this undertaking over the next three years. Late in 1625 a request from the crown to divert money already set aside for the company, in order to finance urgent defense works for Lisbon, drew sharp protest from the câmara, which pointed out the negative effects such interference would have on voluntary subscriptions. Nevertheless, the council did advance the money, which was eventually repaid in government bonds.[24] Then in 1627, after a disastrous Atlantic storm wrecked the two homeward bound Indiamen *Santa Helena* and *São Bartolomeu* as well as seven of the eight warships in the Portuguese home fleet, the Lisbon Câmara was again asked to release money intended for the company to help meet the immediate needs of the carreira da Índia. It provided 20,000 cruzados for the despatch of a carrack and pinnace to

India in April 1627, and a further 40,000 cruzados for three Indiamen sent to Goa in November.[25]

Despite these disruptions the câmara made a firm commitment in December 1628 that it would make available to the company 150,000 cruzados, to be paid in three annual instalments of 50,000 cruzados each. Funds for the first of these were raised from a variety of sources—including a new issue of municipal bonds secured on the *real d'água* (a duty on sales of wine, meat and fish), and even monies previously intended for the purchase of grain—and were delivered to the company piecemeal over the next few months. However, the first instalment was still not fully paid up by the following February when the crown urged the câmara to fulfill its commitment without delay for the money was required to fund preparation of the company squadron due to sail for Goa in 1629. Similar reminders were delivered in subsequent years but it appears that, even if the instalments came in rather slowly, the Lisbon Câmara did eventually pay its full 150,000 cruzados.[26]

Most of the Lisbon Câmara's contribution was delivered in ready cash although a portion, notably that covering the second annual instalment (1629-1630), was in the form of bonds.[27] The câmara was also asked on several occasions to provide the company with emergency loans over and above its original agreed investment. A loan of 4,500 cruzados at 6¼ per cent interest was provided in March 1630, and another was sought in 1631 when the crown urgently requested 20,000 ducats "from any money belonging to this city" for preparation of the company's annual sailings to India.[28] Simultaneously the câmara was making substantial contributions to the viceregal treasury at Goa and also providing cash grants for the war against the Dutch in Pernambuco.[29] Such allocations demonstrate the significance of the câmara as a source of funds for backing officially recognized national causes at the crown's behest.

The Lisbon City Council was not the only câmara expected to lend support to the new trade company. Dom Jorge Mascarenhas had been requested in 1624 to urge the provincial municipalities also to subscribe. He was to give prior assurances that proper accounting and business methods would be scrupulously employed, and that company funds would under no circumstances be diverted to other purposes.[30] Meanwhile the crown also wrote to the provincial câmaras requesting their participation. It followed this up in the new year with the commissioning of a special agent, Dr. Francisco Rebelo Homen, to travel round Portugal "and interest the cities, towns and private individuals in the trade company."[31]

The response to these pleas showed that there was little enthusiasm for the company in the provincial towns of Portugal. Probably typical were the reactions of the câmaras of Vila Viçosa, Fronteira and Borba, each of which expressed loyal support for the company in principle but offered to contribute only 1,000 cruzados a year, over a ten-year period. They claimed they were too impoverished to provide more. In any event the crown was dissatisfied with the attitude of the câmaras and in March 1626 reiterated its demands, asking that contributions be larger and made available without delay. It is not clear whether any significant increase in the câmaras' offerings followed, but in December 1626 the crown announced that it had approved "all that each one of the cities, towns and places of my said kingdoms have offered and will give to the said trading company."[32] At the same time it gave blanket authorization for the councils to raise the funds required from a variety of specified sources. A subscription list of those places in Portugal which did eventually contribute to the company names twenty-eight in addition to the Lisbon Câmara itself. The provincial câmaras provided a total of 168,867 cruzados or slightly more than the 150,000 cruzados invested by Lisbon alone. Three of the provincial contributors—Évora, Castelo Branco and Coimbra—provided 15,000 cruzados or more each; of the rest only Torres Vedras, Beja, Portalegre and Viana do Castelo exceeded 7,500 cruzados.[33]

The crown also tried to squeeze contributions for the company from the câmaras in Portuguese Asia. In 1629 Viceroy Linhares was informed that the crown itself was writing to the Asian câmaras, urging them to subscribe. He was instructed to persuade them to follow the example set by their sister councils at home. The Lisbon Câmara would send its own appeal to Goa, suggesting that the viceregal capital match the Lisbon contribution of 150,000 cruzados.[34] These efforts were almost entirely in vain, for apart from Chaul none of the Asian câmaras offered anything. The explanation given by the Malacca councillors was probably typical. Although "very willing" they simply did not have the money to spare, the city's resources having been used up in the wars against the Achinese of Sumatra.[35] Since only half of the 10,000 xerafins promised by Chaul was ever paid over, the sum total of capital contributed to the company by the câmaras of Portuguese Asia was a miserly 5,000 xerafins (1,875 cruzados), enough for no more than a small fraction of one year's purchases of pepper.[36] The total investment of all municipal bodies, metropolitan, provincial and Asian, amounted to just under 320,000 cruzados, or less than a third of what the crown eventually provided. However, the câmaras' contributions were particularly important because most were made available in cash.

The lukewarm support given by the câmaras made it all the more vital that the company attract private subscribers. On declaring its intention of founding an India company in 1619 the crown had indicated it would seek investment capital from "all persons of whatever quality and status."[37] Such backing would clearly have to come primarily from financiers and merchants in Lisbon, and in 1624 Dom Jorge Mascarenhas was accordingly instructed to interest "men of trade and people who have capital." The Lisbon Câmara urged the crown also to solicit support from the nobility.[38]

Those opposing the formation of the company argued consistently that they saw little prospect of its raising capital from private sources. In July 1625 the two governors of Portugal, the Counts of Basto and Portalegre, advised the crown that no investors would be forthcoming, and late the following year the Council of Portugal asserted that so far not a single businessman had shown interest in the company, and none were likely to do so.[39] These contentions were strongly disputed by Dom Jorge Mascarenhas who claimed that businessmen in Madrid had already indicated their willingness to invest in the company, under certain conditions. If adequate private support was not forthcoming, the crown could simply compel New Christian financiers by decree either to invest directly in the company, or to furnish it with the necessary funds through loans.[40] Early in 1628 Dom Jorge was still expressing optimism about private backing for the company. He claimed that individuals in Madrid had now promised 300,000 cruzados and that this, together with the proceeds of subscription drives then underway in the towns of Castile and the contributions expected from Portugal itself, would make the company "the greatest thing in the world."[41]

Meanwhile, Olivares' "Great Project" of 1624, which had envisaged the creation of a whole series of Iberian companies both to revive trade and to underpin the projected Hapsburg alliance against the Dutch and German Protestants, seemed near realization. On January 24,

1628, Dom Jorge Mascarenhas wrote to the Lisbon Câmara from Madrid that these interlocking companies were now to be founded and would embrace the commerce of Portuguese Asia, the Mediterranean and northern Europe. He affirmed that plans for a company to operate in the Spanish Netherlands and Germany were well advanced. Large squadrons were to be made up in Lübeck, Hamburg and other Hanseatic ports, which would bring badly needed supplies to the Iberian peninsula from the Baltic, and take Portuguese sugar, pepper, salt, tobacco and other commodities to their northern European markets. "This company," he explained, "will work in conjunction with and be dovetailed to that of our India." At the same time, "for the Castilian Indies and the Levant other companies will be formed, and all Italy will go into that of the Levant."[42] As late as June 1628 optimism about these schemes still apparently prevailed in some quarters in Madrid, for an anonymous informant in the Spanish capital wrote to correspondents in Lisbon that "the companies are made and there are four of them."[43]These were the Portuguese India Company, and separate companies for trade respectively with northern Europe, the Spanish Indies (based on Seville), and the Levant (based on Barcelona). Their two leading proponents were reported to be Dom Jorge Mascarenhas and Olivares' Jesuit confessor, Hernando de Salazar.

In fact, by the end of the year Olivares' "Great Project" had been abandoned, and claims that the projected trading companies had already been founded were grossly inaccurate, however likely such a development may have seemed a few months earlier. The reasons for Olivares' change of mind were various. The outbreak of war against France over the Mantuan succession, and the loss of a Spanish silver fleet to the Dutch admiral Piet Heyn in 1682, made even partial crown financing of the "Great Project" extremely difficult. The refusal of the Hanseatic General Council to cooperate in the formation of the projected Northern Company was an equally serious setback which would have crippled the project from the start. Its abandonment was made certain when the lukewarm attitude of Barcelona merchants towards the proposed Levant Company became clear, and when Spanish diplomats failed to reach agreement with the Catholic General Wallenstein for joint operations against the northern Protestants.[44]

The scheme for a Portuguese India company alone survived. It had a compelling raison d'être of its own within or without the framework of a "Great Project," and preparations for its foundation were probably too far advanced by 1628 for it to be easily abandoned. Nevertheless, events proved Dom Jorge Mascarenhas' hopes for substantial private investment in the company as misplaced as his optimism regarding realization of the "Great Project" itself. The 300,000 cruzados promised by business interests in Spain were eventually diverted by order of Olivares to the wars in Flanders, despite the protests of the would-be investors.[45] The company received no subscriptions whatever from Spain, and those from private investors in Portugal were trifling. The official company accounts name only two individual investors: the archbishop of Lisbon, Dom Afonso Furtado de Mendonça, chairman of the board of governors of Portugal in 1626-1630, and Dom António de Almeida, governor of Abrantes, who subscribed 200 and 400 cruzados respectively.[46] No other nobleman, financier or merchant invested in the company either before or after its foundation, and the campaign to attract private capital was therefore almost totally ineffectual.

There was a variety of reasons for this failure. The repeated ill success of the pepper and shipping contracts in the three final decades of the sixteenth century had highlighted the risks and generally poor investment prospects of the official Goa-Lisbon trade, and the reluctance of private enterprise to become involved was notorious by the second quarter of the seventeenth century.[47] Similarly, entrepreneurs involved in trading non-monopoly products on this route were not interested in backing an India company since they already had established channels for the purchase and sale of their merchandise, and were more likely to regard the appearance of the company as a threat than a boon. When a similar company was proposed in 1689 the Portuguese Goa merchants refused to back it because they were already acting as commission agents for private interests in Portugal and feared the loss of this business. Hindu merchants were also uninterested, supposedly because they preferred to keep their money affairs secret.[48]

Good investment terms might have done something to counteract this reluctance. In January 1625 the Lisbon Câmara had proposed that to encourage both corporate and individual interest the crown should guarantee all capital subscribed to the company, but this the crown was apparently unwilling to do.[49] The terms eventually offered were outlined in the company's charter published in August 1628. The minimum subscription was to be 100 cruzados, but persons with less than this sum available could participate by forming groups which would invest in the name of their leading member. Foreigners as well as Portuguese subjects could subscribe. Capital would receive interest at a fixed rate of 4 per cent per annum but could not be withdrawn till expiry of the twelve-year joint-stock, unless after six years the directors decided otherwise. The directors were required to publish the company's accounts at mid-term and could distribute dividends to shareholders once before and once after this point, if the position of the company warranted it. Shares could be freely bought and sold. At the end of the twelfth year final accounts would be prepared, a distribution of capital and profits made and the company dissolved. Subscribers would then have the option of reinvesting in a reconstituted company.[50]

The larger investor was offered inducements in the form of honors, graded according to the size of contribution. Those subscribing 1,000 cruzados would receive the status and privileges of a knight (*cavaleiro fidalgo*) while the wives of those subscribing 4,000 cruzados could use the courtesy title of "Lady." Any investor who provided 30,000 cruzados would receive the rights and privileges of a gentleman of the royal household (*fidalgo da casa del rei*) and anyone who provided a large carrack, fully armed and provisioned, would be given membership in a military order.[51] That these inducements proved insufficient to attract investors was hardly surprising. In late 1626 the Council of Portugal had cited the low interest rate of 4 per cent as a reason for its conviction that the company would be unable to raise capital from private sources. Investors could not be expected to tie up their capital for twelve years, in what was believed to be an increasingly risky trade, for so low a return, especially since they could reportedly earn "over 100 per cent" from investments on the free market.[52] Similar reasons were later given for the lack of response by private interests in Portuguese India. The company board in Goa reported in January 1631 that Portuguese casados in the viceregal capital did not wish to back an enterprise which brought in returns so slowly; they found it more profitable to invest their capital locally where they could get "ten per cent without risk."[53] The verdict of all these potential investors was succinctly expressed by one of Olivares' merchant advisors when he sceptically pointed out that no indication had been given as to how the company proposed to make a profit.[54]

It appears that the crown made no serious attempts actually to coerce or, apart from providing honorific titles and privileges, to induce business interests to come round to backing the company. Dom Jorge Mascarenhas had suggested that New Christians be forced into giving support if necessary, but this proposal was clearly rejected in the company's charter which declared that capital should be subscribed on a strictly voluntary basis.[55] The government was apparently convinced that the long-term interests of the company would be better served if the crown abstained from heavy-handed interference. Moreover, while detailed plans for the company were being completed in the mid-1620's, Olivares was simultaneously conducting negotiations with the Lisbon New Christians for a huge loan of 1,500,000 cruzados for the Iberian Crown. These negotiations were successfully concluded in 1627. They were a part of his strategy to reduce the crown's reliance on Genoese bankers, and probably made it impracticable to demand at the same time forced subscriptions for the company from the predominantly New Christian entrepreneurial elite of Portugal.[56] New Christian investment was further discouraged by the fact that the company's charter explicitly declared subscribed capital liable to confiscation if the subscriber were condemned for heresy.[57] This condition was probably included as a result of Inquisition pressure and was clearly aimed at real or alleged Judaizers who were the principal objects of the Inquisitors' attentions in this period. Significantly, subscriptions to the much better supported Brazil Company of 1649 were exempted from such confiscations.[58]

It was not, in fact, until about 1631 that the uniformly unfavorable reaction to the India Company from merchant and financial interests in both Portugal and Portuguese Asia was fully known. By that time a new Brazilian crisis created by the Dutch invasion of Pernambuco was pressing such demands on the strained resources of Portugal that the government, far from encouraging further investment in the company, was actually considering impounding its pepper stocks to raise funds for a new expedition to Brazil.[59] At about the same time, partly to support the war effort in Pernambuco, the crown extracted another 450,000 cruzados from the Portuguese New Christians. There was therefore little chance in the 1630's that the crown would be willing or able to initiate any moves to channel more funds into the India Company, and the failure to win support from private enterprise was never rectified. Yet the channelling of substantial outside capital into the Lisbon-Goa trade had been one of the major objectives in setting up the company. The failure of such participation to materialize inevitably meant that the company in practice relied much more heavily on the crown than the crown itself had envisaged.

The form and extent of crown support for the India Company was decided in late 1626 after a brief struggle between the Council of Portugal and Dom Jorge Mascarenhas' ad hoc committee.[60] The latter had wanted the crown to provide 500,000 cruzados a year, the estimated amount normally spent on despatching three carracks to India, of which 40,000 cruzados per ship would be a free subsidy and the rest invested capital. The Council of Portugal opposed this largely on the grounds that a net loss for the royal treasury would ensue, but its objections were essentially overruled. At a decisive meeting with Olivares in Madrid on November 22, 1626, attended by Dom Francisco de Bragança, Mendo da Mota and Dom António Pereira of the Council of Portugal, it was agreed that the crown should back the company for a three-year period at the annual rate the committee had suggested, but with the subsidy reduced to 30,000 cruzados per ship. Implementation of this decision was declared to be conditional on adequate prior investment in the company by private business interests. This condition was repeated in the company's inaugural decree issued on August 27, 1628, but of course was never fulfilled.

The inaugural decree also confirmed the crown's commitment to subscribe 1,500,000 cruzados in three annual instalments of 500,000 cruzados each.[61] Part of this would consist of pepper, customs revenue and freight charges from Indiamen expected home in 1628 and 1629, together with these ships themselves, and their stores. Also assigned to the company was the India galleon *Batalha,* which had been detained at Bahia since 1626, and various naval supplies and stores from the royal dockyards, the pine forests of Leiria, the linen factories of Santarém, and other sources. Finally, the company was to receive crown funds that had been earmarked for building ships for the carreira da Índia, the crown's share of profits on the first two voyages under company auspices, and moneys from a number of miscellaneous sources including confiscations by the customs. Dom António de Ataide, count of Castro d'Aire and governor of Portugal in 1631-1633, subsequently claimed that a contribution of 120,000 ducats from the Crown of Castile was also promised, but it was apparently never paid.[62]

The total value of the Crown of Portugal's actual investment in the company reached only about two-thirds of the 1,500,000 cruzados theoretically committed. Dom António de Ataide gave an estimate of just over 1,040,000 cruzados in January 1633, and the company accounts of April 15 of that year indicate the slightly higher figure of 1,056,809 cruzados. The most important single category of items received from the crown consisted of five ocean-going vessels, together with their guns and stores. Only two of these ships, the *Nossa Senhora de Bom Despacho* and the *São Gonçalo,* were in Portuguese home waters at the time the company was founded. Of the others, the *Nossa Senhora do Rosário* had already left for India and did not return until 1629; the *Bom Jesus de Monte Calvário* made the voyage out in 1627, and only reached home again after an exceedingly troubled passage in March 1629; and the galleon *Batalha* from the 1626 fleet reached Lisbon after its time-consuming stopover in Bahia, also in 1629. The value of these five ships, together with about 100 bronze and iron cannon, gun-carriages, gunpowder, shot and small arms, a quantity of masts, spars and other naval stores, and the hulk *São Tomé* which was used as a derrick, was reckoned at 343,217 cruzados.[63]

In addition, the company received the pepper money (*cabedal da pimenta*) on board the *Nossa Senhora de Bom Despacho* and the *São Gonçalo* which had returned to Lisbon in 1628 after unsuccessful attempts to sail to Goa. This amounted to 80,000 cruzados, probably all in silver. To this must be added the pepper, freights and duties brought in by the company's remaining three vessels, which eventually yielded some 371,665 cruzados. The balance of 261,927 cruzados was made up of shipyard expenses credited at Lisbon, and repair facilities made available to two company ships that called in to Luanda. Roughly two-thirds of the crown's actual contribution to the company was therefore in the form of materials, the remainder consisting either of cash or of pepper, freight charges and customs duties, the latter two payable in cash. Most of these assets passed into the hands of the company at a relatively early stage of its existence in 1628 or 1629.

The company's initial resources were thus considerably less than had originally been projected. Its capital was perhaps a little over half what its advocates had hoped, and the shortfall was not made good by any significant investment of fresh funds in subsequent years. Ominously, the Lisbon merchants, "whose judgements, being based on their own self-interest, are always the most reliable," had turned their backs on the project.[64]

6 / The Company Born

In 1628, when Portugal's India Company was founded, Portuguese trade and communications with Asia were administered at the European terminus by the *Casa da Índia* (India House) and the *Armazém da Índia* (India Dockyard). Both these organizations had been in existence for well over a century, but they had recently been placed under the general supervision of a newly created treasury council (*conselho da fazenda*), the principal organ for economic decision-making in Portugal in the years of Hapsburg rule.

The Treasury Council was created by Philip II in 1591 as part of a general program of administrative reform for the kingdom of Portugal. It was composed of two treasury superintendents (*vedores da fazenda*), four councillors, and four secretaries, of whom one was made responsible for matters relating to Portuguese possessions and interests in Asia, Africa and Brazil. The principal function of the council was to exercise broad control over fiscal affairs. It was charged with auditing the accounts of subordinate bodies, and its authorization was required, among other things, for the purchase and provisioning of ships by the Armazém da Índia, and for the various activities, such as the sale of pepper, carried out by the Casa da Índia.[1]

The Casa da Índia had been the body responsible for administering most aspects of Portugal's trade with Asia from the time that direct maritime relations via the Cape of Good Hope had been opened at the end of the fifteenth century. Approximately equivalent to the Spanish *Casa de la Contratación* founded at Seville in 1503, the Casa da Índia received its first detailed standing orders from Manuel I in 1509. However, its earliest predecessor, the *Casa de Ceuta,* was in operation from at least 1434 and the later *Casa de Mina e Guiné,* from which the Casa da Índia evolved, was also of fifteenth-century origin.[2] All imports and exports to and from Asia had to be stored at the Casa da Índia for registration, customs clearance, and payment of freight charges. Its officials also arranged for the sale of pepper and other Eastern products purchased on the crown's account, supervised the loading and unloading of ships, inspected them for contraband, and paid the crews. They also kept registers of all sailings to Asia, and of all individuals travelling on board. At the time of the India Company's foundation the casa's staff consisted of a director, two treasurers, six secretaries, a weight inspector, specialist valuers for gems, pearls and drugs, and a number of guards, porters and other minor personnel.[3]

The naval requirements for the India trade were handled at the Armazém da Índia. This organization was administered quite separately from the Casa da Índia. It had its own director, treasurer and clerks, and also employed a variety of specialized artisans and workmen such as carpenters and sailmakers. In 1628 the director was the experienced Vasco Fernandes César, who had held office for over thirty years, his original appointment dating from 1595.[4] India vessels built in Lisbon were constructed at the dockyard facilities of the Armazém da Índia. Here they were also repaired, fitted out and provisioned. The armazém recruited and supplied crews for the ships, and provided navigational instruments. It functioned under the supervisory authority of the Treasury Council from 1591, until in 1628 control passed to the India Company's board of directors in Lisbon.

Although the Portuguese India Company had been directly inspired by the examples of the Dutch and English East India Companies it was obliged to operate under very different conditions from those which had confronted its rivals. The English and Dutch had no possessions or established interests in Asia when their respective companies were founded. It was therefore possible to concede to the companies not only authority to trade, but sweeping powers to found factories, organize settlements, make war and peace, sign treaties, and govern the Europeans and Asians in their trading posts, through officials appointed by the largely independent boards of directors in London and Amsterdam.[5] The Portuguese, on the other hand, were obliged to graft their company onto a functioning administrative apparatus that had been running an Asian trade empire for a century and a quarter. There could be no question of the new company exercising political and military powers in India since these were already the responsibilities of existing authorities. Even the creation of a Portuguese company, with a strictly commercial role only, required considerable modification of a long-established system of trade administration.

Against this background the Portuguese company was founded by royal decree on August 27, 1628, receiving on the same day a sixty-eight clause charter which set out its constitution, powers and privileges.[6] The charter provided for a board of directors to administer the company's affairs in Lisbon, to formulate company policy, and to supervise a subordinate board in Goa. The Lisbon board, at least in theory, was to be fully independent of all organs of government in Portugal — the entire administrative apparatus in the Portuguese capital, from the Governors of Portugal and the Council of State downwards, being specifically forbidden to interfere in company affairs. Instead, the board of directors was to be directly subordinated to the crown, through the Council of Trade (*Conselho do Comércio*) in Madrid. The charter gave the board the right to submit recommendations directly to this council and further required it to appoint a company representative who would reside at court in Madrid and be entitled to sit as a co-opted member on the Council of Trade when it discussed company matters. On these occasions the company's representative would rank immediately below the most junior member of the council but enjoy full voting rights.[7]

The company board was to consist of a president and six directors, all of whom had to be Portuguese. The president was to be the crown's nominee, while the other board members were to be elected triennially by subscribers holding shares worth 1,000 cruzados or more. As major subscribers the câmaras of Portugal were given the right to choose one of the directors, while the other five were to be elected by the qualifying shareholders in general.[8] These provisions were intended to emphasize the board's substantial independence of official control and influence, but were in practice ineffectual since the lack of subscribers to the company meant that there were no qualified electors apart from the câmaras and the crown. In August 1628 the crown simply nominated all the directors itself "so that the company may be formed immediately," reserving only the câmaras' right to choose their representative.[9] The company board, as then established, consisted of Dom Jorge Mascarenhas as president, Leonardo Fróis, Garcia de Melo, António Gomes da Mata, Francisco Dias Mendes de Brito, and Diogo Rodrigues de Lisboa. The sixth directorship remained vacant since the câmaras never got round to selecting their representative.

Of the five subordinate directors the one most difficult to identify clearly is Leonardo Fróis. He was a gentleman of the royal household (*fidalgo da casa del rey*) and hence a man of some standing, and he had served on the ad hoc committee of 1624. Old and with failing eyesight, he ceased to attend board meetings after April 1629 and was subsequently excused from further service altogether. What qualifications had convinced Olivares and his advisors of Fróis' suitability for board membership are unclear, but he was in all probability an entrepreneur of substance since preoccupation with his personal business affairs was cited as a reason for his withdrawal in 1630, along with poor health.[10]

Garcia de Melo, also a gentleman of the royal household, was a veteran soldier and administrator of long experience in Portuguese Asia, where he seems to have accumulated a sizeable personal fortune. He was vedor da fazenda at Cochin from 1605 until suspended for alleged malpractice in 1610, when a discrepancy of over 1,530 quintals in the annual pepper cargoes despatched to Lisbon was discovered. Although ordered arrested and sent back in the first available ship to Portugal, Garcia de Melo was actually transferred by a sympathetic viceroy to the important and lucrative post of vedor da fazenda at Ormuz where, it was subsequently claimed, he embezzled royal funds. He succeeded in exonerating himself from all charges, although in the Cochin pepper case this was allegedly because of his ability to obtain favorable testimony through bribes and threats rather than proven innocence. Later he was appointed treasurer-general at Goa, the second highest administrative office in the State of India, and was given a China voyage as a reward for good services.[11] On his return to Portugal Garcia de Melo was officially recognized as an expert on Asian affairs and trade, and was regarded by Olivares as an appropriate person to be consulted about the formation of an India company, and ultimately to be appointed a director. However, Melo was an old man by 1628 and died in about December 1630 so that his services were not long available to the company.[12]

The three remaining founder-directors were all substantial merchant-capitalists. António Gomes da Mata was probably a son, and certainly the heir, of a wealthy New Christian called Luís Gomes de Elvas Coronel. The latter had purchased the office of postmaster-general, and in 1607 he was granted a coat of arms and the surname Mata. These honors were subsequently inherited by António Gomes da Mata, who also enjoyed the status of gentleman of the royal household. Apparently he was drafted to the company board because of his wealth and his business expertise and connections. However, his services were also comparatively short-lived, since illness increasingly confined him to his house, and by the end of 1630 his fellow board members felt there was no prospect that he would perform further duties.[13]

The other original directors, Francisco Dias Mendes de Brito and Diogo Rodrigues de Lisboa, both stayed on the board until the company's dissolution in 1633. The former was the eldest son of a celebrated international financier, Heitor Mendes de Brito "the Rich," a man who had often lent money to Portuguese kings. Heir to his father's fortune and business interests, Francisco Dias was a leading figure in Portuguese business circles in the later Hapsburg years, and his cooperation in organizing the company was correspondingly important. Like his father, he was not only a gentleman of the royal household but a member of the prestigious Order of Christ. Later his son, another Heitor Mendes de Brito, was to raise his own regiment to fight for the Braganças, and served with distinction in the Alentejo campaigns, an indication that the family by then was accepted into Portuguese Old Christian society.[14] For despite the attempts of Heitor the elder to prove his descent from a line of knightly forebears traceable back to the early fifteenth century, and a favorable decision of the High Court of Lisbon in 1619 declaring the Mendes de Britos to have been Catholics well before the forced conversion of Jews by King Manuel in 1497, there is little doubt that the family was in fact New Christian. Francisco Dias' younger brother, Nuno Dias Mendes de Brito, was actually one of the principal negotiators who arranged terms with Olivares for the large loan provided for Philip IV by Portuguese New Christians in 1627.[15] Two of his paternal cousins, another Francisco Dias Mendes de Brito and Diogo Mendes de Brito, were arrested by the Lisbon Inquisition in November 1630, and subsequently convicted of Judaizing.[16] The Francisco Dias Mendes de Brito who became a director of the company was also related by marriage to Duarte Gomes Solis, another New Christian entrepreneur who, after a long commercial career in India between 1586 and 1601, had urged the formation of an India company in various memorials and letters in the 1610's and 1620's.[17]

Diogo Rodrigues de Lisboa was a leading member of another Portuguese financier-merchant family, the New Christian Rodrigues of Lisbon. He was arrested by the Lisbon Inquisition in January 1632 on a charge of Judaizing, and the records of his case provide valuable data on his personal history.[18] He was born in 1568 and was therefore already aged sixty when appointed to the India Company board. Widely known as "Crafty" ("*o cantiga*"), he was a Lisbon man born and bred. Although apparently of a somewhat quarrelsome temperament, Crafty was also a long-time friend of the Jesuits of the church of São Roque, to whom he gave generously, and had held office in several religious brotherhoods. He was described by his Inquisition interrogators as a person of not very noble rank, and was actually the only one of the five directors of the company who was not also a gentleman of the royal household. His relatively inferior social status tends to be confirmed by the fact that he was almost invariably listed last on company documents.

However, whatever Crafty lacked in class he made up in initiative and wealth, being accounted extremely rich. Diogo Rodrigues was one of the leading entrepreneurs in Lisbon, a purveyor of loans from the local New Christians to the king in Madrid, and an importer and shipowner of considerable substance who dealt, among other things, in Indian pepper. That he was also remarkably resilient is suggested by the fact that in 1642, ten years after his arrest by the Inquisition and now aged seventy-four, he was farming from the crown the sequestered estates of the pro-Hapsburg Marquis of Porto Seguro, apparently as repayment of loans to the restored Bragança monarch, King John IV.[19] Diogo Rodrigues de Lisboa and Francisco Dias Mendes de Brito, together with Dom Jorge Mascarenhas himself, were the only members of the board of directors of the Portuguese India Company who remained in office for the whole period between 1628 and 1633, and consequently upon their shoulders "all the weight of the business fell."[20]

In order to fill vacancies created by death or incapacity, three additional directors were appointed at later stages of the company's existence. In 1630 an elderly bureaucrat, Luís de Figueiredo Falcão, was brought out of retirement for this purpose. He had had a long and successful career in the royal service, was the author of an important work on the Portuguese national economy and had risen to be a member of the powerful Treasury Council. However, he had already been forcibly retired from the council five years earlier, and was so old and dim of sight that he could not even sign his name. Predictably, Luís de Figueiredo Falcão was of little help to his fellow directors, and his death after a short period made no difference to the work of the board.[21]

The two final draftees to company directorships were Manuel Rodrigues de Elvas and João Salema, and both these men were fit and able enough actually to take part in the board's work. João Salema was described by fellow directors as "a rich and noble man well versed in business matters," but his background remains a mystery.[22] Manuel Rodrigues de Elvas was another New Christian from the Lisbon business community. He was already in his late sixties when appointed to the board, and an old associate both of Diogo Rodrigues de Lisboa, whom he had known for half a century, and of the Mendes de Brito family. Like Nuno Dias Mendes de Brito, he had taken part in negotiations with Olivares for the New Christian loan provided for the crown in 1627.[23]

With the exception of Dom Jorge Mascarenhas, all the directors on the company board were prominent members of the Lisbon business community, although Garcia de Melo and Luís de Figueiredo Falcão had also been important officeholders in the bureaucracy. All but one of the original appointees had the status of gentlemen of the royal household. The majority of directors also came from families involved in one or another of the various commodity and shipping contracts for the Asia trade since the late sixteenth century. Heitor Mendes de Brito, the father of Francisco Dias Mendes de Brito, had been a member of the syndicate which contracted for the pepper monopoly for the five-year period 1592-1597, while the Elvas, Gomes, Rodrigues and Solis families had all figured prominently in similar deals.[24] The government appears to have conscripted to the services of the company men whose families had a background of involvement in the trade with India, and who had accumulated knowledge and experience of its workings.

Under these circumstances it is noteworthy that not one of the directors was also a company shareholder. To ensure his disinterestedness the crown had prohibited Dom Jorge Mascarenhas from investing, yet by contrast expected other directors to do so. At the inception of the company in 1628 the crown declared that those selected as directors who had not invested the minimum 1,000 cruzados were improperly qualified for office. In December 1630 it instructed that any director who had still not contributed should either do so or give up his directorship to someone who would, but to no avail. Actually, there was no provision in the charter specifically requiring directors as opposed to electors to invest in the company, even though critics later apparently assumed this was the case. Of course, the unwillingness to invest in the company displayed by all the directors says little for their confidence in its business prospects.[25]

The company's board in Lisbon was somewhat remarkable for the decrepitude of most of its members. All were elderly, sick, or both—except for Dom Jorge Mascarenhas himself, and perhaps also Francisco Dias Mendes de Brito and João Salema, though it is likely that they too were rather old. A more serious weakness in the view of critics was the board's association with crypto-Jews. Unlike the Brazil Company of 1649, the India Company attracted no capital from New Christians, yet clearly utilized their entrepreneurial skills.[26] Since New Christians were extremely prominent in the Lisbon business community, this was probably inevitable, but it nevertheless gave ammunition to the company's opponents who expressed themselves shocked that an organization financed in the upshot largely by the crown should be run by a group of "Hebrews." Enemies were quick to accuse the directors of gross corruption and inefficiency, especially when the company's affairs began to run into difficulties. The problems of the India trade, it was alleged, arose when control passed from the supposedly honest and untainted hands of those who had served on the Treasury Council to "such a different kind of people as are the Men of the Nation [that is, crypto-Jews] who care for nothing except their own self-interest."[27] It appears that entrepreneurs who served on the Lisbon board probably did have access to certain opportunities for dubious but personally profitable business dealings, as a result of their association with the company. Individual directors were certainly alleged to have fixed contracts for the outfitting and provisioning of ships.[28] But there is no convincing evidence that the Lisbon board was unusually corrupt in its administration of company affairs, and much of the criticism brought against it in this regard smacks of anti-Semitic prejudice. The early seventeenth century was a particularly uncomfortable period for New Christians because of the preoccupation of the Portuguese Inquisition with Judaizing, the reversion of Jewish converts to Judaism. This practice was greatly abhorred in Hapsburg Portugal, and its prevalence probably much exaggerated. New Christians, if condemned as Judaizers, were liable to have their entire estates confiscated and were also harassed by a number of discriminatory measures such as periodic bans on their travelling overseas or selling merchandise abroad. The latter prohibition was re-imposed on the exact day that the crown first announced its intention of founding an India company on February 19, 1619, probably with the intention of trying to channel New Christian capital into the company.[29]

It was largely to obtain a relaxation of such anti-Semitic measures that the Lisbon New Christians had responded to Olivares' overtures in 1621 to provide the Iberian Crown with a large loan. The reluctance of both church and state leaders in Portugal to give concessions to the New Christians probably helped prolong negotiations over this deal. The agreement, concluded in 1627, provided the New Christians with a limited general pardon for sins against the faith, whereby during a three-month period Judaizers could confess, repent and receive forgiveness without further penalty. Restrictions on trade and travel were also lifted, although royal permission to return to Portugal was still required after visiting Protestant countries.[30] Two of the New Christian families involved in arranging this deal—the Mendes de Britos and the Rodrigues de Elvas—were subsequently represented on the board of the India Company. This link probably increased the suspicion with which the company was regarded by those Portuguese who distrusted New Christians on principle and opposed giving concessions to their interests.

The company's prospects of success would certainly have been greater had it obtained the support, or at least approval, of informed public opinion in Portugal. In fact it appears to have been received with either indifference or positive hostility.[31] These attitudes stemmed partly from the conviction that the scheme was economically unsound. The financier-merchant interests in Lisbon clearly thought this, as did the Council of Portugal. There are also indications that the company exacerbated anti-Spanish resentment. Some saw it as an insidious attempt to circumvent the constitutional safeguards given to the Portuguese in the capitulations of Tomar at Philip II's accession. This was apparently the view of Dom António de Ataide, who in 1633 complained that "contrary to the privileges of this Crown, this Company is subordinated to the Council of Trade [in Spain] and not to the ordinary tribunals of this kingdom."[32] Fear of Castilian designs on local Portuguese liberties was strong in the Olivares era, and it intensified the resentment naturally felt by high officials over the company's supposed independence of normal bureaucratic controls.

For the company's well-being it was important that a sound working relationship be established swiftly between the board of directors and the bodies which handled trade and communications with India. With the launching of the new enterprise in August 1628 immediate supervisory authority over the Armazém da Índia passed to the company's board in Lisbon, but the Casa da Índia remained under the supervision of the Treasury Council until 1631.[33] As crown agencies of long standing both these organizations were staffed by royal officials, and a formula intended to safeguard the interests of their personnel, while still giving effective control to the directors of the company, was written into the latter's charter. It provided that the board could make any appointment it wished, but that if the crown officials already serving at the casa and armazém were not kept on, the company must continue paying them their former salaries and perquisites until such time as the crown provided them with equivalent emoluments from other sources.[34] Subsequently, the wastefulness of this compromise was strongly criticized, but the extent to which the company actually did duplicate officeholders in the casa and armazém was probably exaggerated by its critics. Lists of officials employed in both organizations in 1632 suggest that while the company created a few new offices it continued to use serving officials wherever possible.[35] In particular, key appointees such as Vasco Fernandes César, director of the armazém, and Cristovão d'Almada, director of the Casa da Índia, retained their positions under the company's administration.

With the founding of the Portuguese India Company at Lisbon in August 1628 it became necessary to organize a subordinate board of administration in Goa. This task—which could not be accomplished for several months because of the long time-lag in communications—was entrusted to the newly appointed viceroy of Portuguese India, Dom Miguel de Noronha, fourth Count of Linhares and head of one of the great noble houses of Portugal. Born in 1588, Linhares was forty-one when he took up his post in India in October 1629. He was to remain viceroy for just over six years, and his term of office proved one of the most decisive and eventful of the century.

Viceroy Linhares was heir, directly or indirectly, to three distinguished and powerful houses. From his father he had inherited modest lands but a long tradition of service in Portuguese India at the highest level—the Noronhas being one of the handful of families from which viceroys and governors had repeatedly been drawn.[36] At the age of twenty he had also received the lands and titles of the counts of Linhares by the unusual process of nomination, despite being merely a distant cousin of the childless third count, Dom Fernando de Noronha, who, with royal permission, had selected Dom Miguel as his successor.[37] Since this third count of Linhares had married Dona Felippa de

Sá, a sister and heiress of the childless governor-general of Brazil, Mem de Sá, Dom Miguel de Noronha also stood to inherit vast Sá estates and possessions in Brazil.[38]

As viceroy at Goa, Linhares displayed a remarkable capacity for hard work. "I reserve only three out of the twenty-four hours of the day and night for sleeping," he observed in a letter of 1631, and a year later claimed—probably with only slight exaggeration—that he had never worked less than twenty hours a day since his departure from Portugal three years previously.[39] Much of his energy was expended on routine matters. As viceroy, he was obliged to conduct a vast and time-consuming correspondence with his superiors in Europe and subordinates in the Estado da Índia, to grant frequent audiences "because my door is open to everyone at all hours," and to attend to a host of ceremonial duties often lengthy, exhausting and personally costly, from the entertaining of foreign ambassadors to the witnessing of *autos da fé*.[40] A conscientious bureaucrat, Linhares also spent long hours supervising such operations as the fitting out and loading of ships and the embarkation of troops, arguing repeatedly that, because of the inefficiency and untrustworthiness of subordinate officials, these vital tasks could be accomplished in no other way.[41]

Despite this gruelling round of routine duties in a notoriously enervating climate, Linhares showed considerable drive and initiative. His attempts to visit and familiarize himself with the far-flung corners of his vast viceroyalty, even though often frustrated by circumstances, were more in the tradition of the great governors of the early sixteenth century than characteristic of their pedestrian successors of the early seventeenth.[42] In Goa itself he pushed through an important construction programme, providing the viceregal capital with a new gunpowder factory and magazine at Panelim, a military road with a massive stone bridge linking the city with Panjim and the mouth of the Mandovi, a new hospital (the *Hospital de Piedade*) and improved fortifications.[43] Dogged continuously by Asian hostility and Dutch and English competition on the one hand and by chronic shortages of money, men, munitions and ships on the other, not to mention the ravages of famine, Linhares also did well to contain the military crises that confronted his administration, especially in Sri Lanka, Malacca and Mombasa. Most important, however, was his realization that Portugal could no longer sustain war against all her European trade rivals in the Indian Ocean, and his decision in consequence to arrange in 1635 a truce with the English East India Company at Surat, thereby initiating in the region a new Portuguese policy that was to last for over three centuries.

These developments lay well in the future when Viceroy Linhares left the Tagus for Goa in the spring of 1629, carrying with him precise instructions for establishing the company in India.[44] He was to set up a local board of administration composed of five members to handle company affairs in Asia. The main responsibility of this body would be to organize and administer, under the general instructions of the company board in Lisbon, the provisioning, repair and fitting-out of ships, the sale and purchase of merchandise, and in general all matters concerned with the receipt and despatch of the company's exports and imports.

The leading appointee to the original company board in India was Manuel de Moraes Supico, a native of the Portuguese province of Trás-os-Montes, who had migrated to India as a young man at about the turn of the century. The social status of his family is impossible to determine with certainty, but whatever it may have been it did not prevent his business activities in the Orient from attaining a success which made him perhaps the richest Portuguese merchant in Goa.[45] He seems to have taken an early interest in the Africa trade, since he was acting as a factor in Mozambique in about 1613 and was operating Goa-Mozambique voyages thirteen years later. He was also active in the import-export trade with Lisbon and was one of the chief members of a group of speculators who contracted for the Macao-Nagasaki voyages over a three-year period from 1629, for the record sum of 306,000 xerafins. He made loans to the government for use in such widely separated places as Muscat and Malacca, acted as treasurer for the *collecta,* and dealt in the diamond trade.[46]

That Manuel de Moraes Supico's business activities achieved outstanding success is demonstrated by the honors and offices he had attained by the time of his death. A gentleman of the royal household, he became in 1629 a knight commander of the Order of Christ. He was also a councillor of the Goa Câmara and, in 1629-1630, served as president of the misericórdia. His wealth ensured his family's acceptance into society in the viceregal capital, and his daughter married as her second husband António de Sousa Continho, a governor of India. One of his sons married the daughter of a vedor da fazenda at Goa, José Pinto Pereira, who was later John IV's ambassador to Sweden, while a granddaughter married the son of a viceroy, Dom Rodrigo Lobo da Silveira.[47] Manuel de Moraes Supico's personal success story is briefly related on his simple but dignified tomb in one of the finest private chapels in the cathedral of Goa. His death in May 1630, only a year after company operations had begun in India, must have been a severe blow to the Goa board.[48]

The Goa company director "on whom fell the greater part of the work" was Francisco Tinoco de Carvalho.[49] He was a member of an extensive entrepreneurial family with business interests in Lisbon, Madrid and Seville, as well as in Goa and the Portuguese Asian region. Tinoco de Carvalho was particularly active in the Goa-Lisbon trade, and in 1630 exported to Europe a large quantity of cotton piece-goods, as well as a variety of other products, such as wax, indigo, cardamom, cinnamon and fans. He was a tax-farmer, and on at least one occasion contracted, in partnership with another merchant, to outfit ships for the China voyage. In 1628 he was acting as commercial agent for Philip IV's queen, Elizabeth of Bourbon, buying pepper for her in Kanara and arranging for its shipment to Lisbon, and he was factor in India for the Monastery of the Incarnation at Madrid. Francisco Tinoco de Carvalho also showed interest in the trade in cowries, making a contract for their purchase with the ruler of the Maldive Islands in 1628 and selling surplus stock to the company for export to Lisbon the following year.[50] Eventually, with the winding up of the India Company in 1633, he returned to Lisbon. In 1638 he was apparently still engaged in the intercontinental trade between Europe and India, but had shifted his base of operations to Madrid.[51]

Concerning the remaining founder-directors of the company board in Goa—Fernão Rodrigues de Elvas, Fernão Jorge da Silveira and Valentim Garcia—details are much hazier. Fernão Rodrigues de Elvas was presumably a member of the New Christian Rodrigues de Elvas family of Lisbon. If so, he would have been related to Manuel Rodrigues de Elvas, one of the company's directors in Portugal. Like Manuel de Moraes Supico he died within a year of his appointment, in June 1630.[52] Fernão Jorge da Silveira, who became the company's treasurer in India, was also involved in exporting Asian products to Lisbon as a private merchant. Although he was reportedly very ill in January 1631, he was still a director of the company when it was finally dissolved in 1633.[53] Viceroy Linhares once described him as "a man of very low birth, and a Jew," though this was written in a moment of exasperation.[54] Fernão Jorge da Silveira was in fact a member of one of Portugal's most prominent New Christian families—a brother of Pedro de Baeça, the wealthy Lisbon capitalist.[55] Little is known about Valentim Garcia except that he remained on the board until the company's liquidation, participated in the cloth and other commodity trades from Goa to Lisbon, and also engaged in tax-contracting.[56]

When these five nominees were summoned before Viceroy Linhares the day following his disembarkation at Goa in October 1629, none was at all pleased at the appointment thrust upon him. All complained that to accept responsibility for running the affairs of the company in Goa would involve them in serious inconvenience and personal loss, and Linhares later reported to the crown that he had had much difficulty in persuading them to comply. Nor did the directors become reconciled to their appointments with the passage of time. In 1631 they all asked to be released from further service, claiming that their initial three-year terms were nearly up and that some of their number wished to return to Portugal. They also claimed to be either sick or too preoccupied with other essential business to be able to devote the time required to attend to the company's affairs. However, the Lisbon directors and the crown were unimpressed by these pleas, and the Goa board, with the exception of changes made necessary by deaths, and the belated co-option of vedor da fazenda José Pinto Pereira in 1632, retained its membership unaltered until liquidation.[57]

In mid-1630 Manoel Correia and Bartolomeu Sanches Correia were appointed to the board to replace the two deceased members. Manoel Correia's background is obscure, but he was described in 1631 as very old and unable to do much work.[58] More is known of the chequered career of Bartolomeu Sanches Correia. This man was a wealthy New Christian merchant whose interests were sufficiently large to make it possible and worthwhile for him to cultivate the patronage of the great. His contacts included the Count-Duke of Olivares, to whom he sent the present of an oriental desk in 1625, and the king's confessor, António de Sotomaior, to whom he sent a praying-stool. Shortly before the arrival of the Count of Linhares at Goa in 1629, Bartolomeu Sanches Correia had become farmer of the Goa customs.[59] He attempted to use this fact to get excused from service on the company board, and in January 1631 the Goa directors wrote to Lisbon that he could "with very just cause" be permitted to withdraw, since the customs took up all his time, making it impossible for him to attend to the affairs of the company.[60] Linhares was later rebuked by the crown for supposedly relieving this tax-farmer of his directorship, but it appears that the viceroy had in fact not acceded to his request. Sanches Correia was subsequently involved in a prolonged dispute concerning alleged frauds over the customs contract, although he appears eventually to have been exonerated from any misdoings.[61]

The reluctance of all the Goa directors to be involved in the running of the Portuguese India Company is understandable in the light of the problems and difficulties it brought them. Administering the company's affairs meant expending time and effort on a scheme which offered them little or nothing personally, and which was felt to have dubious prospects of success. Moreover, the company's freedom of action in Goa was in practice even more limited by the power of established government agencies than it was in Lisbon. Defense measures, the provision of dockyard and warehouse facilities, the despatch of the annual fleet to Lisbon, and all major decisions on general commercial policy depended on the local viceregal authorities without whose cooperation the company was virtually powerless. The crown had instructed that the directors in Goa be given every assistance, but in fact the viceroy and his principal subordinates, who had many other heavy responsibilities and extremely limited resources, were often in no position to give the backing required. Further, the directors were forced to rely on the cooperation of crown officials in India, who resented the company's existence and saw it as a threat to their vested interests.

If the viceroy failed to give the company adequate support, the only appeal possible was to the court at Madrid. This entailed a delay of from eighteen months to two years before even a tentative decision was received, which was in turn enforceable only by the local viceregal authorities. The company's complete subordination to the government in Goa was demonstrated from the first: immediately after his arrival in 1629, Viceroy Linhares, faced with an empty treasury and the need to despatch the Malabar pepper fleet without delay, took a forced loan from the company's supply of silver specie that had arrived on the carracks from Lisbon, despite the supposed inviolability of these funds, as trading capital.[62] Such dependence did not augur well for the Portuguese India Company as, in the early autumn of 1629, its newly appointed directors turned their attention to the commercial operations of the coming year.

7 / The Goa-Lisbon Trade under Company Administration

The presence of established trading interests in the Portuguese Empire at the time of the India Company's formation made it necessary to define carefully the company's field of operations. In the early planning stages it was apparently envisaged that the company would trade not only in the State of India but also in Mina and Guinea, where presumably it would have participated in the slave trade. This suggestion was later dropped, and the charter issued in August 1628 merely stated that the company's objective was to revive the Portugal-India trade, omitting all reference to West Africa.[1] The company directors in Lisbon did not take this as necessarily confining them to the Lisbon-Goa route. In February 1630 they proposed that a trade link be established between Lisbon and Macao, via Mozambique and Malacca. This would have bypassed Goa, normally the compulsory entrepôt for all Portuguese seaborne commerce between Europe and Asia. The board suggested that initially two 300-ton vessels be used for this voyage, and that on arrival at Macao they be made available for freighting by the captain of the Japan voyage and the local Macaonese merchants, for the passage on to Nagasaki.[2]

This was an obvious attempt to break into one of the few highly profitable trades remaining to the Portuguese in Asia. It met a predictably lukewarm reception from the Treasury Council in Lisbon, and was firmly opposed by Viceroy Linhares and his council at Goa. Linhares pointed out that small fast galliots were then used by the Portuguese operating in and near the Straits of Malacca to evade Dutch patrols. The larger vessels contemplated by the company would be too slow and vulnerable for this purpose. More important from Linhares' viewpoint, direct trade with Macao would reduce customs returns at Goa and create unwelcome competition for local merchants engaged in the trades between Goa and the Far East.[3]

The threat of such competition alarmed the Goa Câmara, which would certainly have endorsed the warning of an official two decades earlier that "the China trade is the principal activity on which the Portuguese in India exist, and without it the state would be destitute."[4]

The câmara strongly opposed conceding to the company "the right to send ships to Malacca, Manila, and China," emphasizing that the Goa merchants were more dependent than ever on their rôle as middlemen in the Europe-China trade.[5] This widespread opposition to the company's proposal apparently had its effect. The scheme was not accepted, although the possibility of direct communications between Lisbon and the Far East by various routes continued to be discussed in Portugal for several years afterwards.[6] In the meantime, the company was obliged to concentrate almost exclusively on trade between Portugal and the west coast of India, with a few minor exceptions such as the saltpeter trade with Bengal.

Careful consideration was also given to the nature and extent of the company's monopolies on the Goa-Lisbon route. A fundamental decision was needed as to whether the company should receive, like the English and Dutch East India Companies, a complete monopoly on all trade with India, or whether it should merely take over the commodity monopolies previously reserved to the crown, and compete on equal terms with private merchants for other products. Olivares himself sought advice on this question from experts such as Garcia de Melo and Ferdinand Kron, the latter a highly experienced business agent who had been resident factor in Cochin and Goa for the Welsers and Fuggers for almost forty years (1587-1624). Kron thought both alternatives unworkable. He doubted that the company could find sufficient capital to operate a complete monopoly. Were such a monopoly granted he believed it would bring a drastic decline in the Lisbon-Goa trade, for the Portuguese inter-port traders at Goa would cease to function as intermediaries, and Asian merchants would no longer bring their stocks to Goa before the arrival of the annual carracks from Lisbon. On the other hand, the company could not compete successfully in a free market against local entrepreneurs, who would be able to undercut its business by utilizing their established contacts in Gujarat and elsewhere.[7]

Olivares found such scepticism exasperating. Rather unreasonably, he informed Kron that his arguments were contradictory and instructed the old merchant to reconsider, concluding with the characteristically blunt and egotistical demand, "You have pointed out the difficulties—for love of myself, try to overcome them."[8] Whether Olivares received the more positive advice he sought is unclear, but when the company was eventually founded it received only certain specified commodity monopolies, plus the right to collect freight charges and customs duties on private merchandise shipped aboard its vessels. The monopoly commodities granted were coral, ebony, cowries and above all pepper. The possibility that the company might later receive monopolies on the export of other products, especially cinnamon, was left open by the charter.[9]

It was the responsibility of the India Company's board in Lisbon to provide the necessary funds for the purchase of pepper and other Asian commodities, the refitting and provisioning of the company's ships at Goa, and any other expenses incurred by the Goa directors on the company's behalf. Almost all these monies had to be shipped out from Lisbon in the form of bullion or specie, as had long been the case under crown administration of the official India trade. In 1629, 98 per cent of the company's working capital in India was provided in this way, and a similar overwhelming reliance on funds from Portugal was maintained throughout the company's existence.

The silver and gold shipped to India from Portugal for official trade purposes was generally known as the pepper money (*cabedal da pimenta*) since usually it was all, or practically all, consigned to purchase pepper. Half a century before the company's foundation, Linschoten had declared that "the most and greatest ware that is commonly sent into India, are [*sic*] rials of eight," and this position remained substantially unchanged into the first half of the seventeenth century.[10] Although there were significant quantities of gold in each of the consignments of pepper money brought out by the company ships to Goa, silver in the form of Spanish reals-of-eight invariably predominated. Silver reals accounted for 62 per cent of the total shipment of specie and bullion to Goa in 1629 and nearly 68 per cent in 1630. For 1631 the exact proportion is unclear, but in 1632 and 1633 it appears that the company's pepper money consisted exclusively of silver. Gold was despatched to India in significant quantities by the company in its first two years (Appendix 2.4). A small fraction was in four-cruzado gold pieces, but most was uncoined since gold produced a higher rate of profit when sold on the Goa market in this form. In 1629 coined and uncoined gold together made up more than a third of the year's pepper money; in 1630, just over 20 per cent.[11]

On the arrival of the annual fleet in India, the silver and gold was promptly unloaded and stored for safe-keeping in the Franciscan convent at Goa. Much of the silver was subsequently sold off to merchants and money dealers from Goa and from the neighboring Indian states, the latter flocking to the viceregal capital for the occasion, their purses "full of gold pardaos." Since the late 1560's the demand for imported silver—especially in Gujarat and other parts of India, and in China—had been consistently strong, so that the Portuguese had no difficulty in selling their stocks at good profits. The viceregal authorities attempted to restrict the steady outflow from Goa in the early seventeenth century but with little success.[12] Profits on the sale of silver in Goa fluctuated between about 25 and 40 per cent during the early and mid-Hapsburg periods, but the general tendency was upwards. When the Portuguese India Company commenced operations the rate had increased to over 50 per cent. In 1629 the company's gross profit on the sale of its silver reals was 70 per cent, and although this decreased to just over 53 per cent in 1630, sales remained highly lucrative (Appendix 2.4).

By contrast, gold was generally worth more in Europe than in India in the sixteenth century and appears to have been seldom exported in that period. However, by the mid-1620's gold as well as silver was often despatched to Goa on the carracks. In 1627 as much as 32,847 out of 40,000 cruzados that comprised the year's pepper money was in gold—a proportion of over 80 per cent. In that year uncoined gold was being exchanged in Goa at a more favorable rate than silver.[13] Later, under the company regime, Portuguese gold was sold at rates of profit comparable to those of silver, and the directors continued to despatch both metals whenever possible. In 1629 the sale of gold four-cruzado pieces and uncoined gold brought a gross profit of 62 per cent, and in 1630 it brought 56 per cent. The average profit on combined gold and silver imports to Goa in the years 1629-1631 was 58.4 per cent, and remained at over 50 per cent for at least the next few years.[14] Goa was recognized as an important outlet for gold at this time, and even agents of the English East India Company sometimes sought to exchange their supplies there rather than at Surat, in the expectation of better returns.[15]

It is apparent that silver reals were not used directly to buy pepper in Malabar and Kanara. Pepper was normally paid for by the Portuguese either in local silver coinages as specified in treaties with the rulers of the pepper ports, or in gold.[16] In the time of the company the gold *São Tomé,* a coin struck at the mint in Goa since 1548, was commonly used in Malabar. São Tomés were coined with gold obtained from a variety of sources beyond the Cape, particularly Mozambique, as well as from the gold bars brought out on the carracks, when these were available. Sometimes funds were needed too urgently to wait for the coining of gold shipped from Portugal, as in 1631 when the company directors at Goa borrowed 40,000 xerafins from the local misericórdia because "there was little time after the arrival of the carracks, and we did not have enough [time] to mint the gold which came."[17] On the other hand, the directors emphasized that "for Cochin São Tomés are the best money there is," and the company accounts for 1629-30 show that these coins were much used to pay for pepper at Quilon also.[18] The gold São Tomé was likewise highly regarded by the English for use in the India spice trade, and in 1635 an English East India Company representative bound for Goa was instructed to have his gold coined into "St Tomaes" to pay for pepper.[19]

Nevertheless, in the late 1620's and early 1630's it seems normally to have been more profitable to sell gold by weight on the open market at Goa than to coin it into São Tomés, largely because of the costs and dues payable at the mint and the high prices for gold offered by Hindu and Moslem dealers. In 1627 gold coined at the *Casa da Moeda* yielded only 197 xerafins 28 réis per mark compared with 202-204 xerafins per mark for uncoined gold sold to merchants and bankers. In 1629 the gap had narrowed somewhat, but uncoined gold at Goa still brought a higher return. For the purchase of Kanara pepper the Portuguese usually had to obtain gold Vijayanagar *pagodas* which were required for transactions with the nayaks of Ikkeri.[20] Sometimes uncoined gold was transferred directly from Goa to the pepper ports, but this was likely to involve a reduction in profits owing to the lower value of gold sold in this form in Malabar and Kanara. In 1629 a mark of gold which was worth 196 xerafins 4 tangas at Goa rated only 185 xerafins 2 tangas at Cochin and 176 xerafins 2 tangas 30 réis at Quilon.[21]

Apart from gold and silver the only European commodity that the company sent to Goa in any quantity was coral, which it shipped out every year except in 1629 and possibly 1631. Coral composed 13 per cent of the value of company exports from Lisbon to India in 1630, and equalled about a third of the value of gold and silver shipped in 1633.[22] The company obtained its supplies in Italy and maintained a factor at Pisa. In 1631 he was responsible for despatching sixty-nine cases of coral from Livorno to Lisbon, which the company paid for partly in pepper and partly in bills of exchange. The following year it purchased coral from a certain Francisco Bresano.[23] In marked contrast to the consistently large profits gained from the sale of silver and gold, the Portuguese India Company found the market for coral somewhat unpredictable. In 1630 the Goa directors wrote home to Lisbon that there was then little demand in India, partly because of political turmoil in Ikkeri following the death of Venkatapa Nayak in 1629, and partly because of competition from the Dutch at Masulipatam. The following year the market was again weak, and little more than a quarter of the year's consignment was sold. Nevertheless, the company persisted in its efforts to sell coral and in both 1632 and 1633 made good sales, and profits.[24]

From time to time the Portuguese India Company also exported to Goa on its own account some non-monopoly general cargo goods including mercury, English cloth, cochineal and small arms, and sold there surplus stores and provisions from its Indiamen.[25] Although a variety of European goods—jewellery, certain textiles, foods and wine—were then commonly despatched to India by private Portuguese merchants, the company appears to have taken no significant interest in these products.

For those who administered the Portuguese pepper trade, making funds available in India at the most convenient time of year involved persistent and apparently unresolvable difficulties. Pepper supplies purchased early in the season were more reliable, their quality better, and their cost less than those bought late. Also, if the pepper were bought and transported to Goa in good time, the carracks could be loaded so as to begin their return voyage by the end of January, thereby reducing the chances of shipwreck, or long delays at Mozambique, caused by the winter storms at the Cape of Good Hope. The difficulty was that while the pepper had to be bought if possible between late February and late May, the carracks which brought the pepper money from Portugal did not normally reach Goa until August or September.

The logical solution, which the Goa directors themselves recommended, was for the board in Lisbon to provide Goa with sufficient funds to carry over from season to season. This had been both urged and attempted before, and in 1612 the new standing orders had specified that one half of the pepper money should always be shipped out to Goa a year in advance. The Goa directors indicated in 1630 that the sum required would be at least 60,000 cruzados annually. However, the company in Lisbon could not or would not provide this money so far in advance, any more than the crown had done earlier. The Goa directors were therefore left to manage as best they could on existing resources, and on such expedients as came to hand.[26] Since they could not normally purchase pepper on company credit they were forced, when cash was lacking, to rely on loans, repaid on the arrival of the ships from Portugal.

In the initial year of 1629-30 no loans were necessary. However, in 1630-31 sufficient money was borrowed from local financiers to buy, before the arrival of the carracks, 5,000 quintals of pepper. This was secured on the personal guarantees of the directors since "as officials of the company there is no one who would entrust us with a penny."[27] The company paid 10 per cent interest on its loans, the highest rate permitted by the regulations of 1612. Loans raised in this manner were usually insufficient to avoid completely the need to buy some pepper after the arrival of the carracks from September onwards. This could result in hasty and dangerous last-minute loading of Indiamen for the return voyage. Nevertheless, despite all the well-known weaknesses of the system, when the crown again took over the Lisbon-Goa trade in 1633, it reiterated that the circumstances of the royal treasury made it impossible to despatch funds for the pepper in advance, and brushed aside the Count of Linhares' warning that such advance provision was essential if the ships were to get away on time.[28]

The relatively little pepper money exported to India between 1629 and 1633 confirms that the company's financial resources were too meagre to achieve the growth in the Goa-Lisbon trade that proponents had hoped for. The average annual value of company cargoes sent from Lisbon to Goa during these years was 71,899 cruzados. In a comparable period of five years at the start of the Hapsburg era (1580-1584) the contractors for the pepper monopoly had agreed to despatch 151,800 cruzados annually—over twice as much as that achieved by the company. In the eighteen years preceding the company's foundation the pepper money reaching Goa had averaged 96,222 cruzados per year, still well above the company's average.[29]

Other factors besides reduced pepper money contributed to the company's trading difficulties. Business was to some extent hindered by shortages in the supply of pepper, as of other Indian commodities, caused by famine. In 1630, although the Goa directors were able to report that the carracks were returning with only a little vacant cargo space, they also complained about high prices and the general lack of all kinds of goods. In 1631 they described the state of the neighboring country as miserable, adding that they did not know how they could fill the ships. In that year one hold of the *Santo Inácio de Loiola* remained empty, and another partly so. In 1632, although the only vessel despatched to Lisbon was the state pinnace *Nossa Senhora dos Remédios,* conditions were so bad that even this ship could not be provided with a full cargo, and freight charges had to be lowered as an inducement to private customers.[30] Only in the final year (1633) does an appreciable improvement in supply conditions appear to have occurred, the viceroy describing the cargoes of that year as "very rich" and made up of "goods of high price."[31] Unfortunately, by then the company was past redemption.

Though shortages of both funds and goods for purchase obviously limited the volume of company business, it is unlikely that the directors would have followed a boldly expansionist policy even had they possessed the resources to do so. In 1632 the Lisbon board informed the crown that it would be against company interests to send five rather than three carracks to fetch cargoes from Goa: with the participation of other European powers in the Asia-Europe trade, the market would become oversupplied and profits correspondingly decline, were the Portuguese to increase their turnover substantially.[32] This may have been partly a rationalization of the company's inability to despatch more than three vessels in one year, but was in any case a discouraging indicator for its prospects.

By far the most important of the products exported by the Portuguese India Company from Asia was pepper. Pepper accounted for over 81 per cent of the company's exports to Portugal in 1630 and 96 per cent in 1631. Thus it is obvious that the company's fortunes depended overwhelmingly on the profits of this single commodity.[33] For four out of the five years between 1630 and 1634, when carracks were available at Goa for the voyage back to Europe, the company despatched an average of 11,050 light quintals (9,669 heavy quintals) of pepper to Lisbon.* The exception was 1632, when only a small government pinnace returned to Portugal, and the company despatched

*Unless otherwise stated, a "quintal" of pepper will denote a "light quintal," equivalent in weight to 87.5% of a heavy quintal.

no ships or pepper at all. This reduced the annual average under the company regime to 8,840 quintals. Consignments actually reaching Lisbon were invariably much lower. During the period 1628-1632 an average of only 5,479 quintals per year arrived in Lisbon, or about five-eighths of the average quantity loaded annually at Goa between 1629 and 1633 (Appendix 2.2). The discrepancies between sizes of cargoes despatched from India and those arriving in Portugal are attributable chiefly to shipwrecks, delays of shipping en route, and the losses in weight which normally occurred in pepper cargoes transported to Europe. The extent of the discrepancies obviously contributed to the company's financial difficulties by 1633.

The export performance by the company compared most unfavorably with that achieved up to the first decade of the seventeenth century. Figueiredo Falcão assumed an annual intake of 20,000 quintals of Asian pepper at Lisbon in the 1590's, and postulated the standing orders for the pepper trade in 1612 as 20,000-24,000 quintals per year out of India.[34] These figures may be too high, but they do indicate the level of imports which well-informed contemporaries expected at the turn of the century. A more realistic approximation of the volume of shipments actually achieved in the sixteenth century can be obtained by averaging the figures for pepper imports at Lisbon, which are known for about thirty of the years between 1501 and 1598.[35] This suggests that, on average, 16,873 quintals per year reached Lisbon from India during the century, a level which implies about 18,000-19,000 quintals out of Goa, if allowance is made for losses from shipwreck and other causes.

Thus the quantities of pepper exported by the Portuguese India Company were down to probably about 50 per cent of the average level despatched from Goa in the previous century. At the same time, the amount of pepper reaching Lisbon under the company was down to between one-third and one-half of its sixteenth-century rate. However, this record reflects on the company less adversely if compared with the results achieved under crown monopoly in the second and third decades of the seventeenth century, when average annual totals of pepper exported from Goa to Lisbon were already down to 10,355 quintals per annum (Appendix 2.2).

The company normally purchased pepper in the traditional export centres of Cochin and Quilon on the Malabar coast and Onor, Barcelore and other smaller factories on the coast of Kanara. It also occasionally bought from agents at Goa, especially in years when supplies and time were short. Small quantities of pepper originating in Quilon were bought from five Portuguese traders at Goa in 1630, and three years later Viceroy Linhares reported that Ramaqueny, "the principal merchant of the mainland, very rich," had sold the company "great quantities" of pepper, which he brought from the producing areas at his own cost. The Portuguese preferred when possible to seek pepper from both the Malabar and the Kanara ports, but from at least the second decade of the seventeenth century they usually bought most of their supplies in Kanara. Between 1612 and 1629, twice as much pepper was exported from Kanara to Lisbon as from Malabar. Although the India Company actually purchased 83 per cent of its first annual pepper exports (1630) in Malabar, this was a temporary expedient made necessary by political differences with Vira Bhadra Nayak in Kanara. By 1633, the company had reverted to buying its entire stock in Kanara, and concentration on this source was maintained in 1634 (Appendix 2.2).[36]

Pepper prices in Kanara were consistently higher than those in Malabar and, despite some fluctuations from year to year, showed an increase which amounted to between 30 and 40 per cent for the quarter-century 1610-1635. In Malabar prices remained fairly stable over most of this period, but then rose somewhat in the early 1630's from 10½ to 14 xerafins per heavy quintal. The main reason was a falling-off of supplies from the growing areas, which the company's factor in Cochin attributed to the high prices then being offered for Malabar pepper on the Coromandel coast. There the Dutch were established at Masulipatam, and there was a local trade in pepper with Bengal and other areas to the north. It was largely as a result of this, and of a new, more favorable agreement with Vira Bhadra, that the viceroy was instructed in 1634 to buy as much pepper as practicable in Kanara, though still maintaining Portuguese interests in the Malabar market as far as possible.[37]

As a consequence of the rise of Ikkeri the Portuguese had been forced to purchase most of their Kanara pepper from the nayaks at relatively high contract prices, although supplies above a stipulated quota could be bought from private traders at the normal (and lower) market price. An agreement to this effect had been made by Viceroy Fernão de Albuquerque (1619-1622), with the result that, by 1623, Portuguese ships "every year fetch pepper out of his [Venkatapa

Nayak's] dominions, and bring him in a great sum of money."[38] For the 1626 season a new contract provided that the Portuguese buy 600 *candy* of pepper from the nayak at 28 pagodas the candy, and a further 300-400 candy at 24 pagodas, if there were sufficient funds. In 1632 the Portuguese had an arrangement to buy 500 candy of pepper from the nayak at 26 pagodas a candy, but Linhares was pressing negotiations for a further contract, which was eventually agreed upon in April 1633. The Portuguese were thereby required to buy only 350 candy of pepper annually from Vira Bhadra Nayak at a reduced price of only 22 pagodas a candy, while supplementary supplies bought from private traders were expected to cost as little as 20 pagodas the candy. However, unfortunately for the Portuguese, when the 1634 buying season began Vira Bhadra proved unwilling to adhere to the agreed price and demanded that it be restored to 28 pagodas the candy. Naturally the viceroy was reluctant to accede to this, and what the Portuguese saw as the unreliability of Vira Bhadra emphasized the importance to them of continuing to make regular purchases of pepper in Malabar as well as Kanara. At about the same time Vira Bhadra granted permission to the English to found a factory at Bhatkal, a development which much alarmed the Portuguese although the scheme was abandoned by the English soon afterwards.[39]

The success of the Portuguese pepper trade was dependent as much on selling prices in Europe as on buying prices in India. The table of pepper prices prevailing at the Casa da Índia during and immediately before the company's years of operation, however incomplete, suggests that in Lisbon these prices were probably fairly stable from about 1620 through the early 1630's, but substantially lower than in the previous two or three decades.[40] In fact, by the time the India Company began operations prices were regularly well below the 32 cruzados a quintal which in the late sixteenth century was the established minimum price acceptable at the Casa da Índia (Table 6). This, combined with the simultaneous trend for pepper prices to rise in the principal Indian supply areas, especially in Malabar in the 1630's, confirms the crown's comment in 1634 that pepper prices in Cochin were "so high" while in Portugal they were "so much reduced."[41]

It is apparent, then, that both the quantity of pepper exported and the margin of profit attainable were considerably less under the company régime than they had been for most of the sixteenth century and the first two decades of the seventeenth, and that the trend was for

Table 6. Selling Price of Pepper at Lisbon (in cruzados per quintal)

Year	Amount
1590's	30
1615	34-31 ½
1617	45-42 ½
1621	28
1627	19 ½ - 17
1628	22 ½ - 19
1629	22
1630	25
1631	24
1632	24

Sources: Figueiredo Falcão, *Livro,* p.6; AHU, codex 1164, ff.22v-23, 54v, 109v-110; Lynch, ff.62, 62v, 167v-168, 242; BN Rio, codex Pernambuco 1/2-35, f.139; De Silva, "The Portuguese East India Company," pp. 185, 203.

conditions to become increasingly tight. Nevertheless, in the 1620's and 1630's the pepper trade could still be profitable provided the bulk of the cargo leaving Goa actually reached Lisbon. This is demonstrated by the successes achieved by the two company carracks which reached Europe without any significant losses in 1630 and 1631. The first of these, the *Santíssimo Sacramento,* loaded pepper bought for 16,532 milréis in India, and sold in Lisbon for 40,148 milréis, a gross profit of 143 per cent. The pepper cargo of the second vessel, the *Bom Jesus de Monte Calvário,* fetched 153 per cent gross profit when sold in Lisbon.[42]

These healthy profit margins, which were hardly excessive given the nature and conditions of the trade, indicate what could be accomplished by the company only under the most advantageous circumstances. Unfortunately for the company, it was favored with such circumstances only in the cases of the two ships mentioned above. Of its five large carracks that left Goa for Lisbon before 1633, three failed to arrive intact: the *São Gonçalo* was a total loss—and both the *Nossa Senhora de Bom Despacho* and *Santo Inácio de Loiola* suffered substantial losses of cargo, though the latter ship at least still showed a relatively small gross profit of 30 per cent. The abortive voyages of the India carracks which left Lisbon in 1631 and 1632, and the fact that the company was unable in consequence to despatch any vessel at all from Goa in 1632, considerably exacerbated its difficulties. An organization with substantial capital resources and determined backing might have been able to overcome these setbacks, especially in view of the encouraging performances of the *Santíssimo Sacramento* and the *Bom Jesus de Monte Calvário*. But for the Portuguese India Company, which had so little in reserve, a few major failures, together with an increasingly unfavorable price situation, threatened disaster.

In terms of value, products other than pepper made up about 20 per cent of the Portuguese India Company's exports from Goa to Lisbon in 1630, 4 per cent in 1631, and only an insignificant proportion in the remaining years. Most of these subsidiary exports—indigo, ebony, cowrie shells and rice—were handled purely for commercial profit. However, there was also a second category of products, consisting of saltpeter, and various ships' parts made from tropical hardwoods, which were exported for naval or military rather than commercial reasons, and for these the maintenance of the supply was more important than the size of the profit, if any.

Indigo, widely used as a blue dye by European clothmakers from the early sixteenth century, was the most important of the company's secondary commercial exports. Accurately described by Linschoten as "a costly colour, and much caryed and trafiqued into Portingall," indigo was exported mainly through Surat and Cambay on the northwest coast of India.[43] It was widely grown in the northern part of the subcontinent, with Sarkhej in Gujarat and Biana near Agra the principal producing centres. In the early seventeenth century both the English and the Dutch considered indigo still to be an important Indian export commodity.[44] Indigo accounted for 11½ per cent of the total value of the Portuguese India Company's exports in 1630. For that year the company bought 325 quintals from a Gujarati trader at 60 xerafins the quintal, and paid a 7 per cent customs duty at Goa. The cost of leather and gunnysacks for packing, the labor of making up the packages, the transportation and other handling costs added a further 2-3 per cent. The sale of this indigo, which was shipped on board the two carracks which did eventually get back to Lisbon, brought a good return: 126 per cent gross profit on the *Santíssimo Sacramento* consignment, and 116 per cent on that of the *Nossa Senhora de Bom Despacho*. The Goa directors realized the importance of the indigo trade. In 1630 they suggested asking the crown for a monopoly, but in fact the company was unable to send home any further consignments. The immediate reason was supply shortage owing to the famine conditions which developed in northern India in late 1630 and which also affected the English indigo trade at Surat.[45] But even had the famine not occurred it is unlikely that Indian indigo would have retained its highly profitable position in the European market much longer. Already in the 1630's it was being ousted by American-grown indigo with which the Asian product could not compete either in quality or in price.[46]

Rice, another profitable export commodity, was also hurt by the famine. In 1630, 1,500 sacks of rice had been loaded into the three homewardbound carracks, filling holds for which pepper was not available. Although the entire cargo of the *São Gonçalo* disappeared with the ship, and the rice on board the *Nossa Senhora de Bom Despacho* had to be jettisoned, that from the *Santíssimo Sacramento* fetched a 253 per cent gross profit when sold in Lisbon, the best rate achieved on any of the company's exports for which records are available.[47] Despite its high profitability, rice was too bulky a product to ship home in quantity large enough to make a significant contribution to the company's overall trade position. It was unsuited to the Goa-Lisbon trade which, with its high freight costs and limited transportation resources, had to concentrate on luxuries. The rice carried home in the *Santíssimo Sacramento* in 1630 made up less than 1 per cent of the total value of the company's exports to Lisbon that year. The following year, with famine intensifying, Indian rice was much dearer and in shorter supply, and the Goa directors informed Lisbon that no significant quantities could be loaded for Europe.[48] As far as can be ascertained, no more rice was exported by the company in the remaining years of its existence.

The only other trade commodities which the Portuguese India Company exported from India were cowrie shells and ebony, neither of which was supplied from the subcontinent. Cowries were obtained from the Maldives, the patchwork archipelago of coral islands and islets in the shallow waters of the western Indian Ocean, while ebony was acquired through the Portuguese trading posts in Mozambique in East Africa.[49] There were reliable markets for cowries in Bengal and Siam, which provided outlets for Portuguese country traders.[50] However, the shells were also shipped to Europe—under a company monopoly inherited from the crown—for eventual re-export to West Africa. There they were used as currency in the slave-trades of Guinea and Angola where "one cannot make payments satisfactorily without cowries."[51] In 1629 the Goa directors were instructed to send home up to 2,000 quintals of cowries in the next fleet, and consignments were in fact despatched on all three carracks in 1630. About three-quarters of the total was purchased directly from the ruler of the Maldives; the rest was provided by one of the company directors, Francisco Tinoco de Carvalho, who had previously made a contract with the Maldivians. However, these cowrie cargoes were worth so little on arrival in Portugal that the Lisbon board instructed Goa to suspend further shipments, and no further consignments were sent while the company lasted.[52]

Mozambique ebony, a prized timber for furniture and carvings, became an established component of the Portuguese Indian Ocean trade in the sixteenth century.[53] It was the least important of the commercial products shipped by the company to Europe in 1630, and comprised less than 1 per cent of the total value of exports. Yet it was the only commodity, apart from pepper and saltpeter, that the company continued to buy and ship home in subsequent years. Ebony was a company monopoly, and in its first year of trading the Lisbon directors had instructed Goa to send home 200 quintals in each carrack, a request that was impossible to fulfill. Lisbon had also suggested that the Goa directors come to an agreement with a Mozambique contractor to provide regular supplies, but this also proved impossible. In practice the company appears to have bought its ebony from various middlemen in Goa who must have acquired it, directly or indirectly, from contacts in Africa. In 1630 the Goa directors bought 297 quintals of Mozambique ebony from Simão Cardozo, Diogo de Lucena, and Father Miguel de Paz for over 1,415 xerafins, or 4¾ xerafins the quintal.[54] The following year both the *Santo Inácio de Loiola* and the *Bom Jesus de Monte Calvário* carried small cargoes of ebony, while in 1632 Viceroy Linhares extracted permission from the directors to send home a consignment of high quality ebony in the government pinnace *Nossa Senhora dos Remédios.* Although the ebony was to be divided equally between the Countess of Linhares and the company, the crown took this as a breach of the company's monopoly and expressed its severe disapproval. Despite the persistence of the company's dealings in ebony this product was no more successful than cowries. Most of the consignments of 1630 and 1631 remained unsold in the Casa da Índia in 1633.[55] Thus, apart from dealing in small quantities of indigo and rice in its first year, the company's trade in subsidiary commercial products from the Indian Ocean area proved a rather miserable failure.

Saltpeter, which was in short supply in Europe, was a long-established component of Portugal's India trade and had been brought back to Lisbon aboard the carracks at least since the reign of John III (1521-1557). The almost yearly repetition, from the late sixteenth century onwards, of urgent royal requests for as much of this commodity as possible indicates the strength and persistence of the demand.[56] At its foundation the Portuguese India Company was made responsible for the saltpeter trade to Portugal and was enjoined, in the customary manner, to despatch as much as possible. However, in 1630 it managed to send only 168 quintals, while the two company Indiamen of 1631 carried only 95 quintals between them. The following year no company ships made the return voyage to Lisbon, but the viceregal authorities consigned 200 quintals of refined saltpeter aboard the pinnace *Nossa Senhora dos Remédios,* on the crown's account. When company sailings from India to Portugal were resumed in 1633, the Goa directors were able to send a mere 57 quintals. Company shipments of saltpeter made up about 1 to 2 per cent of the total value of its exports in any one year.[57]

The Goa directors' poor achievement in the saltpeter trade was partly an inevitable product of the scattered and unreliable nature of supply. The company purchased some of its saltpeter at the west coast ports of Barcelore, Mangalore and Rajapur, though the supplies originated beyond the Ghats. However, the most important outlets were in Coromandel and Bengal, both regions well away from the company's main sphere of activities. Additional supplies were obtained in Sind in the far northwest.[58] The India Company never really succeeded in organizing satisfactory procedures for obtaining saltpeter. At first (1630) its purchases were made from local Portuguese and Eurasian contractors at the various west coast outlets—from Domingos Carreiro, for example, a casado living at Barcelore, and from Fernão Carvalho at Cochin, who handled Bengal saltpeter. Later, in 1631, the Goa directors put forward a scheme which would have eliminated some of these middlemen. The proposal was that the company take control of the local inter-port trade, sending pepper from the west coast of India to sell in Bengal and elsewhere, in exchange for saltpeter. However, if this scheme was ever approved it was not effectively actuated, and in 1632 the company was still trying to organize saltpeter purchases through a contractor.[59]

The difficulties regarding saltpeter were aggravated by competition from the viceregal authorities both for the limited supplies available to the Portuguese in India and for use of the refining facilities at Goa. Although high quality saltpeter was occasionally shipped to Portugal without further purification, it was more usual to refine stocks at Goa in advance. For this the company required the services of the new gunpowder plant constructed at Panelim on Tissuary Island by Viceroy Linhares.[60] At crown insistence the viceregal government somewhat reluctantly granted the company these services, though reserving priority for its own needs.[61] In practice this meant that the company seldom had access to the plant, which was almost always in use for the Estado da Índia, and even when the company did succeed in getting its saltpeter refined it was—if the directors are to be believed—badly shortchanged. In the company accounts for 1629 a payment of 50 xerafins to the master of the gunpowder plant was recorded, "for the work he had with this saltpeter and to make him favourable to our needs in this business."[62] Despite greasing this particular palm, in 1631 the directors felt obliged to report to Lisbon their astonishment that from a consignment of 160 quintals of crude saltpeter, little more than 30 quintals of the refined product had been obtained. A year later the viceroy was moved to issue a denial that the company had been swindled by the gunpowder factory.[63] Ironically, even if the company had succeeded in overcoming supply and refining difficulties it would have gained nothing from the saltpeter trade to Portugal since the crown, which purchased such consignments as reached Lisbon, apparently paid only cost price (Appendix 2.5 and 2.8). This was despite the fact that the selling price in Lisbon was sufficiently high to allow a profit of well over 200 percent on the free market.[64]

In addition to saltpeter, the company despatched to Europe a certain amount of naval equipment, also on a non-commercial basis. The excellent quality of Indian timber for ship-building had been recognized by the Portuguese from the sixteenth century, and carracks constructed wholly in Indian shipyards had served in the Lisbon-Goa voyage since at least the 1550's.[65] The Lisbon board of the company was, therefore, acting in a well-established tradition when it asked the Goa directors to supply certain naval necessities. In response to this request, the carracks of 1630 carried back to Lisbon a variety of items including mast-tops, pumps, capstans, rudder-shafts, and blocks.[66] Timber for this equipment came from Cochin and was comprised of teak, poon and other local hardwoods.

In addition to its own goods, the company shipped home aboard its Indiamen each year a considerable quantity of private merchandise, on which it levied freight and certain duties. The income obtained shows how important were these private shipments. It received 23,212 milréis on the private cargo aboard the *Santíssimo Sacramento* in 1630, a total which made up about 33 per cent of the gross return to the company on the ship's voyage from Goa back to Lisbon. The proportion for the *Nossa Senhora de Bom Despacho,* which reached Lisbon in 1631, was about 27 per cent while on the *Bom Jesus de Monte Calvário* the proportion was again in excess of 30 per cent.[67] Thus duties and freight charges were of more value to the company than any other item in the Goa-Lisbon trade, with the exception of the pepper sales.

Although merchandise freighted home on private accounts is nowhere detailed in the available company records, two outside sources partially fill this gap. These are an apparently complete list, drawn up on the orders of Viceroy Linhares, of items which passed through the Goa customs house for despatch to Europe on the three carracks of 1630, and a shorter summary of private cargoes carried on the homeward voyage of the *Bom Jesus de Monte Calvário* in 1631. The customs list of 1630, which included 112 different commodities, indicates that the private cargoes that year were made up chiefly of a wide range of Indian textiles.[68] Though some of these had reached Goa overland through Bijapur, and others had been brought by sea from Bengal and Coromandel, the overwhelming majority probably came from Gujarat on the coastal fleet. In addition to textiles, the cargoes included a variety of spices, among them cinnamon, cloves, ginger, camphor, cardamom and musk. There was also a selection of oriental finished products such as gilt caskets, finely worked furniture, screens and carpets. The names of over two hundred persons freighting these goods are listed, including the viceroy and several priests. All the names are Portuguese, indicating that those involved were Europeans, Lusitanized Eurasians or Catholic Indians. It is possible that Hindu merchants also participated through Christian front-men—though this is nowhere stated and seems unlikely, since the Banyans, while willing to despatch merchandise on crown vessels, apparently considered doing so on those of the company to be too risky.[69]

The cargo list for the *Bom Jesus de Monte Calvário* in 1631 divided private shipments into three sections covering textiles, spices and gems. Again, textiles appear to have comprised easily the largest part of the cargo, though only the numbers of cases and bales were noted, without further details. The spices section itemized each product separately showing that wax, benzoin, incense, indigo, ginger, long pepper, cloves and senna were among the commodities shipped. The gems consisted of pearls and *bizalhos,* the latter probably meaning small bags of precious or semi-precious stones.[70]

Also carried aboard the company's ships, wholly or partially duty-free, were the traditional allowances (*liberdades*) of the crew members, and certain specified quantities of pepper or other goods, conceded by the crown to a few privileged individuals and institutions. In 1630 the flagship *Nossa Senhora de Bom Despacho* loaded exports in the latter category for Dom Francisco da Gama, Count of Vidigueira, the Monastery of the Incarnation at Madrid, and the Captain-General Dom Jorge de Almeida. The viceroy despatched some cargo in the *Santíssimo Sacramento* on which the Goa board had requested that he be excused freight. Concessional cargoes belonging to important or influential owners of this kind were usually shipped in the hold for drugs (*paiol das drogas*). Other small private cargoes were placed in *gasalhados*—particular compartments most of which, at least in the first instance, were assigned to the crew. The majority of these compartments were available on the middle deck and consisted of one or more divisions formed by the ribs of the ship's hull. Others were in the steerage area on the upper deck. Compartments which were not already the perquisite of a crew member were supposed to be sold by the company authorities in Goa. In 1630 their sale yielded 1,275 xerafins, most of the purchasers being ship's officers such as the master and the pilot, who wished to supplement their perquisites. The sale of such spaces on the other two carracks of that year increased this total to 1,585 xerafins. Income from this source was not very significant, therefore, but most of it was put to good use as the company earmarked it to pay cash bonuses to seamen who had been unable to fill their liberty chests.[71]

The shipping of private cargoes involved frequent disputes and difficulties during the period of company control. In 1630 there were too few compartments available for the large number of sailors, gunners, dependents of the officers, and servants, whose rights and claims had to be met. According to the Goa directors this was largely because many compartments had been sold in advance in Portugal to private individuals. When the crews therefore found the space available for their liberty-chests reduced, they became mutinous.[72] To prevent such disturbances the Goa board urgently asked for a provision forbidding anyone from buying compartments from the authorities in Lisbon and excusing its officials from respecting such sales if they did occur.

A predictable result of such situations was dangerous overloading of the ships. One of the directors who visited the flagship *Nossa Senhora de Bom Despacho* shortly before her departure for Lisbon in 1630 found that there was insufficient room for the crew's liberty-chests and that the ship was loaded to capacity, with thirty-five persons still lacking places. The ship's officers later blamed the company for having overloaded her with rice and for selling "places it was not customary to sell."[73] She subsequently struggled into Lisbon with the greatest of difficulty while one of her consorts, the *São Gonçalo,* was wrecked en route. In 1631 and again in 1632 the problem was not overloading but a shortage of private cargoes, aggravated by the deepening famine. Returns on freight and duties indicate that some non-company merchandise was shipped in those years; but nevertheless the carracks sailed home loaded far below capacity.[74] On the *Bom Jesus de Monte Calvário* in 1631 the only spaces sold were two steerage compartments. In order to fill the ships for their return voyages in lean years the company directors at Goa suggested that Banyans should despatch goods to Lisbon, just as they sent merchandise to the various Portuguese settlements and trading posts in India. The crown instructed the viceroy to encourage this development, but unfortunately the Banyans so distrusted the company that they refused to freight anything on vessels it controlled, though they indicated they would be willing to do so if the carreira da Índia were run by the crown.[75]

For much of the sixteenth century cloves, nutmeg and mace had accounted for an important portion of Portugal's oriental trade, but this position had decisively changed with the expulsion of the Portuguese from Ternate in 1570, and their loss of Amboina and Tidore to the Dutch in 1605. Thereafter only an unreliable trickle of Indonesian spices found its way to Goa. During the company regime small quantities of cloves and other East Indies spices were shipped out on the carracks of 1630 and 1631, as private cargoes for individual merchants.

Similarly, the company shipped Sri Lanka cinnamon to Portugal on behalf of private merchants rather than for itself, despite the great importance of this commodity and the express wishes of the directors for monopoly rights.

Cinnamon from Sri Lanka had begun to play a much bigger role in Portuguese Asiatic trade by the last decade of the sixteenth century, making up to some extent for the cloves, nutmeg and mace of Indonesia to which Goa had only limited access after 1570.[76] Although the Portuguese often sold part of their cinnamon to Asian buyers, most of it appears to have been shipped to Lisbon on the carracks, partly on the crown's account and partly on that of private individuals. It was the favorite spice for making up the duty-free allowances which each member of an Indiaman's crew was permitted to ship home, and its presence in quantity was very evident on many of the homeward-bound vessels early in the century, sometimes cluttering up the decks and endangering stability at sea. Jean Mocquet left Goa in January 1610, on a carrack which could only sail "with a great deal of trouble because the ship had cinnamon almost as far as the middle of the mast."[77]

In the period immediately before 1619 an average of about 2,000 quintals of cinnamon is said to have been carried on each carrack sailing from Goa to Lisbon, and the trade continued to flourish into the 1620's.[78] However, at the beginning of the 1630's exports were severely disrupted as a result of the hostilities which broke out between the Portuguese at Colombo and Raja Sinha II of Kandy, and which climaxed in a major disaster for the Estado da Índia with the defeat and death of the Portuguese commander, Constantino de Sá de Noronha, at the hands of the Sinhalese on August 20, 1630. This was to be followed by over a quarter of a century of bitter warfare and varying fortunes until the Portuguese were finally expelled from the whole island of Sri Lanka by the Dutch and Sinhalese in 1658.[79]

Ironically, only seven months before the defeat of Constantino de Sá, the company board in Goa had requested the crown for a monopoly of the lucrative cinnamon trade to Lisbon. The directors proposed that a strict limit of 3,000 quintals of cinnamon should be brought each year from Sri Lanka to Goa, exclusively on the crown's account. This would then be purchased by the company, at an agreed price (which the Goa directors felt would be better arranged in Lisbon than in Goa, since the viceroy and crown officials in India could be expected to impose a far harder bargain). Once purchased by the company at Goa, the cinnamon would be despatched to Portugal in amounts of not more than 1,000-1,500 quintals per carrack, and would be properly stowed aboard. This would have the advantage of avoiding much of the prevailing confusion and inefficiency in loading, with unlimited amounts of cinnamon brought aboard in the form of private shipments and allowances. It would also provide the company with a considerable profit.[80]

Apparently the Portuguese India Company was never granted the cinnamon monopoly. This was probably because of opposition to the idea from both the Treasury Council in Lisbon and Viceroy Linhares at Goa, who argued that such a grant would run counter to royal interests.[81] Yet, the accident of war apart, the company's expectations for cinnamon were not unreasonable. It had been yielding higher rates of profit than pepper, while the large quantities handled in some years of the early seventeenth century showed that its value must on occasion have equalled approximately half that of the pepper exported. After the debacle against Kandy in 1630 and the consequent curtailment in cinnamon supplies, the trade recovered at least temporarily, and in 1633 relatively large quantities were again reaching Goa from Sri Lanka. In 1635 a cargo of about 5,560 quintals was expected to reach Goa.[82]

The increased importance of cinnamon in Portuguese trade in Asia is one of the factors which explain why Portugal strove so tenaciously to maintain its loosening grip on Colombo and its hinterlands between 1630 and 1658. Cinnamon also helps to explain why proposals to move the viceregal capital from Goa to Sri Lanka were seriously broached on at least two occasions in the Hapsburg period.[83] It was ironical that one of the few lucrative lines the India Company might have developed remained, despite the directors' pleas, outside its control, while it was forced at the same time to concentrate on such relatively unpromising products as cowries, ebony and saltpeter.

8 / Company Shipping

If the Portuguese India Company were to have any real prospect of success, then reasonably efficient operation of the shipping service (*carreira da Índia*) between Lisbon and Goa was an obvious necessity. In a general sense this was the case for whatever authority handled Portugal's trade with Asia, but most particularly for the company, which risked failure and liquidation if even a few voyages were lost.

During the 1629-1634 period, the company owned and operated nine large carracks. It built three of these itself in the Lisbon shipyards, and also completed a fourth—the *Santíssimo Sacramento*—which had been under construction when the company was formed. The other five ships, all in more or less operational condition, were donated by the crown in 1628 as part of its official investment.[1] The company's Lisbon directors apparently believed in using large carracks in the carreira da Índia rather than those of small or medium size which were recommended by most Iberian naval experts of the day.[2] Of the three carracks built wholly under company auspices in Lisbon, at least two—the *Nossa Senhora de Belém* and the *Nossa Senhora da Saúde*—were large vessels with four levels or decks. Contemporaries were particularly impressed by the massive size of the *Nossa Senhora de Belém,* a vessel "of great capacity," described by her own captain on the fatal voyage of 1635 as "the most beautiful, best-built, and largest ship that ever sailed this voyage"—"a mountain of wood."[3] Similarly, most of the vessels inherited from the crown appear to have been large, even by Portuguese standards.[4]

No Indiamen were built by the company in India itself, but it is clear that the Goa board shared the views of its Lisbon counterpart on the matter of size. In 1630, and again in 1631, the Goa directors suggested that resources be made available to build one carrack in India each year, recommending that it have four decks. This suggestion was never carried out, although Viceroy Linhares undertook at Goa the construction of two large galleons to be used for the royal service in Asian waters. Linhares, who wished his galleons to be bigger and better than their predecessors, planned their capacities at about 700 tons and their armaments at fifty to sixty cannon apiece. A subsequent request from the Goa board that these vessels be sold to the company as India carracks at cost price was rejected, and the company was obliged therefore to rely wholly on shipping provided by the directors at Lisbon.[5]

The company's nine India carracks together served a total of thirty-five years on the Lisbon-Goa route. Their average length of service was three years eleven months each, during which most of them completed less than two return voyages between Lisbon and Goa. This record was marginally better than that of the Portuguese Indiamen of the preceding decade (1618-1627) which lasted on the average only three years three months. It was about the same as the overall record achieved by the carracks and galleons in the preceding forty years during which time the mean was also three years eleven months, but for slightly fewer return voyages (Appendix 3).

These figures clearly say more for the dismally poor achievement of Portuguese Indiamen in the Hapsburg years in general than they do for any improvement under the company. The company's barely better than average performance was largely the result of favorable sailing conditions in 1629 and 1630 and did not reflect superior organization. Most contemporaries, from the sailors who manned the company ships to the Hindu merchants at Goa who pointedly declined to consign cargoes aboard them, regarded the company's handling of the carreira da Índia as less efficient than that of the crown. More specifically, for every squadron of Indiamen, except that of 1633, which left the Tagus for Goa under the company regime, vehement and detailed complaints were made against the Lisbon directors for alleged maladministration. These criticisms came mainly from those who actually sailed on the ships, and were at least in part confirmed by the company's board at Goa.

According to the Goa board, the company's first ships to sail from Lisbon to India, which brought with them the new viceroy, the Count of Linhares, in 1629, departed Europe before repairs and fitting-out had been properly completed. Consequently on their arrival in Goa in October that year these ships were in extremely poor condition. They had also been badly provisioned and most of their vital sardine supply, allegedly already rotten when brought aboard at Lisbon, had had to be jettisoned.[6] Under the circumstances it was understandable that Viceroy Linhares had been compelled to put into Mozambique for extra food and water, after much sickness and many deaths among his men. Since it was the responsibility of the company to have its ships provisioned, Linhares felt justified in ordering payment to be made for these emergency supplies with money taken from the pepper funds aboard the *São Gonçalo.*[7]

In defense of their handling of the carracks on this occasion, the Lisbon directors pointed out that their board had not been formally constituted until October 1628. By that time work on the three Indiamen was already woefully behind schedule, and on the two older vessels—the *Nossa Senhora de Bom Despacho* and the *São Gonçalo*—it had not even begun. Under the circumstances it was to their credit that the ships had been ready by March 1629. The directors also affirmed that an official inspection of the vessels just before embarkation showed that they were perfectly prepared and in fact "more generously and better provided with supplies and other necessaries than for many years." Moreover, these company ships did at least reach Goa whereas two of the six accompanying royal galleons were lost en route.[8]

If the company board at Lisbon could reasonably disclaim some of the responsibility for the shortcomings of the 1629 sailings, it had much less convincing excuses for 1630, when the situation appears to have been even worse. In that year ships reached India in only five months and ten to twelve days, a much better than average passage. Yet they were so short of provisions and naval stores on reaching Indian waters that the viceroy ordered the local board to send them "some refreshments, cows, and other things" while they were still twenty leagues out of Goa.[9] Within a month of their arrival at the capital, and following complaints from the ships' officers and an official report from the Goa directors, Linhares decided to order a full inquiry into the conditions on board this fleet. This he entrusted to a judge of the Goa High Court, Luís Mergalhão Borges, before whom proceedings began on November 13, 1630.[10] Oral depositions from thirty persons who had come out on these ships, ranging from the captain-major, Dom Jorge de Almeida, to teenage cabin-boys, were taken down at these hearings. Nearly all witnesses were strongly critical of conditions on board, and most of those who had made previous voyages to India compared the company administration unfavorably with that of the crown.

The most common complaint was against the poor quality and inadequate quantity of provisions. Dom Jorge de Almeida declared that the biscuit was very bad, that the cod provided only one meal—the rest of the supply proved so putrid that it had to be thrown overboard—and complaints abounded concerning the poor quality of the wine. He added that if the voyage had lasted the normal six months or longer many men would have died from hunger. The experienced admiral, Cristovão Borges Corte-Real, who commanded the *Bom Jesus de Monte Calvário,* spoke in similar terms, claiming that the biscuit was the worst he had ever tasted, the wine was vinegar, the rice was inadequate, and there were no onions, "it being a very important matter to be supplied with rice, onions and other things of this kind, in hot lands."*

Other witnesses testified that for the last month of the voyage there was no wine left at all for those dependent on the ships' general supplies. Confirming the views of their commanders, the vast majority of other witnesses claimed that the biscuit and such wine as there was were far below standard in quality, although a second pilot and two other men considered them more or less normal. There were no dissentients to the general opinion that the rations were inadequate in quantity, and almost half of those interviewed affirmed that there would have been deaths from starvation had the voyage lasted much longer.

Another major complaint concerned the shortage of essential naval stores on board the two carracks and the alleged rottenness of much of the spare equipment they did carry. "The officers were crying out for tackle," said one sailor, and inadequate or rotten shrouds and stays were declared by almost half of the witnesses questioned to be a serious handicap to the proper working of the ships. Speaking of conditions aboard the *Bom Jesus de Monte Calvário,* the vice-admiral remarked that he had frequently asked his ship's master for nails, planks and sailcloth during the course of the voyage, only to be told that "the company had not given them to him, not even so necessary a thing as planks." On the *Santo Inácio de Loiola* Dom Jorge de Almeida's pilot had complained to the admiral that he had never seen cables so white with rot before. Only two of those interrogated contended that the tackle was of good quality.

Many of the witnesses also alleged bribery and corruption in the manning and provisioning of the two ships. One such accusation charged members of Dom Jorge Mascarenhas' family of having been

*Although onions are less rich in vitamin C than many other fruits and vegetables, they contain enough to be of some value against scurvy.

bribed by a Lisbon vintner to purchase unsaleable vinegar as wine for the carracks, but most complaints of this nature concerned recruitment of the officers and crews. It was commonly rumored in both ships that certain officers had bought their places on board from Dom Jorge Mascarenhas, his wife or his sons. The master and pilot of the *Santo Inácio de Loiola* were particularly singled out for this accusation. Similar charges were made regarding some of the sailors, and Cristovão Borges Corte-Real related how he had heard one experienced hand protest in the Casa da Índia that he had not been signed on because he did not have the money with which to bribe Dom Jorge Mascarenhas.

It was further claimed by a number of those questioned that a large proportion of the so-called seamen and gunners recruited by company officials in Lisbon were actually tailors and cobblers, with no experience of the sea or of the handling of guns. Some witnesses, including Cristovão Borges Corte-Real, specified that thirty-five out of the forty gunners on the *Bom Jesus de Monte Calvário* were greenhorns. The vice-admiral added that he had protested to the company board before leaving the Tagus that four of his ship's twenty-four cannons were missing. Thus, poorly fitted out, improperly provisioned, badly manned and inadequately armed, the two ships, in the views of many of their crews and passengers, had escaped disaster only through the mercy of God and an exceptional run of good weather. Linhares warned in a covering letter to the crown that, even though all the accusations made at the inquiry had not been proved, sufficient evidence had emerged to show that the situation was serious and stern royal action was required.

Despite the scandalous conditions aboard the company's outward-bound ships in 1629 and 1630, all of them completed safe and relatively speedy passages to Goa. Then, in the pivotal year of 1631, fortunes changed. The two carracks despatched to India—the *Nossa Senhora de Belém* and the *Nossa Senhora do Rosário*—were forced back to Lisbon from off Pernambuco in mid-September. Fitted out again, together with the *Santíssimo Sacramento,* in 1632 they had to abandon the voyage even before clearing the Tagus. These failures were serious setbacks for the company. They brought it, in the words of its treasurer, "a very great loss" on the first occasion and "a very considerable loss" on the second—the latter despite, and perhaps even partly because of, the hasty despatch of three smaller vessels more suited to out-of-season sailing, as emergency substitutes.[11] The regular ocean-going carracks finally made the passage to Goa in 1633, but this was too late to benefit the company, which had been liquidated while the fleet was still on the water.

Faria e Sousa blamed the abortive voyage of 1632 on the fact that the ships could not get out of the Tagus because of constant bad weather, but did not account satisfactorily for the fiasco of 1631.[12] In fact, more than the elements was involved in these cases. António de Saldanha, the admiral on both occasions, claimed that in 1631, when the crown's instructions required that the carracks be fully prepared for sea by February, they were not actually ready until April 18, the day they weighed anchor from Lisbon. Moreover, although regulations required that for a projected six-month journey ships should carry one-third extra rations to cover possible emergencies, these vessels were insufficiently provisioned for even the bare six months. Saldanha had complained to the governor of Portugal about this, and orders were given that the normal rations be supplied; yet still nothing was done. Saldanha claimed that he had attempted the voyage in 1631 reluctantly and only out of deference to the king. Within a few days of departure he realized that the ships were not carrying enough provisions for even four months, that most of the biscuit was rotten, and that most of the fish, which had to be thrown overboard, was likewise bad. Almost everyone on the flagship fell sick, less than forty persons remaining wholly fit. Saldanha was forced to put his men on half-rations, but even this extreme measure failed to prevent the abandonment of the voyage "as a result of lack of provisions which would not have happened if the carracks were supplied as they should have been."[13]

Regarding the 1632 case Saldanha affirmed that, although the crown had been informed that the ships were properly prepared and had been unable to leave Lisbon merely because of adverse weather, the truth was that among the bad days there were some fair ones on which the voyage could have been commenced. Shortages, he argued, were every bit as bad as in the previous year. Even though an investigation ordered by Dom António de Ataide as governor of Portugal had confirmed this state of affairs and ordered it corrected, nothing was actually done. Pointing out that some essential dockyard supplies—including timber—had to come from overseas, he complained that only the smallest quantity and worst quality were made available for the

India carracks, the rest being siphoned off for other uses. He also reiterated the now chronic complaint of inadequate artillery, saying that although the ships were supposed to carry a minimum of thirty guns apiece, the flagship had received only twenty-six, and the other two vessels twenty-two and twenty-three respectively. Some of these cannon were sub-regulation six or eight-pounders, despite the fact that encounters with enemies were now increasingly likely.

Saldanha also charged that positions on his three ships had been purchased by bribery, though he did not actually specify to whom the bribes were paid. Instead of recruiting genuine seamen and trained gunners, the authorities at the Casa da Índia had allowed "cobblers, potters and any dirty rabble" to buy their way onto the fleet, and even the purser and quartermaster aboard the flagship had purchased their offices. The crews came to be composed mainly of "youths of small age, all greenhorns" who lacked discipline and military experience. There was not a single physician or surgeon on the whole fleet, although each carrack did have its barber. Since the Portuguese home fleet, which had relatively easy access to Iberian ports, had its own medical officers, Saldanha argued that the India carracks, with their much more arduous voyage and normally no ports-of-call en route, should be similarly provided.[14]

Naturally Dom Jorge Mascarenhas as chairman of the company board in Lisbon did not accept the validity of these charges and criticisms. It is also obvious that many of the complaints made against the company's naval administration concerned conditions endemic to the early seventeenth-century India fleets in general. Nevertheless, the company clearly failed to achieve any noticeable improvement in this administration.

Of a large sampling of sixty carracks and galleons which served on the carreira da Índia during the forty-year period before the company's foundation, twenty-eight were eventually wrecked, and five fell victim to European or Algerine enemies (Appendix 3). This amounted to a remarkably high attrition rate of 55 per cent, with the losses distributed about evenly between the outward voyage (sixteen) and the homeward voyage (seventeen). Under the company regime between 1629 and 1633 no ships were actually lost on the outward voyage, but three out of nine carracks were wrecked on the homeward voyage. Therefore, while the company performed better than the crown had been able to do on the outward voyage, it achieved a slightly worse record on the return voyage with 33 per cent losses as against 28 per cent.

This relatively poor showing for the Goa-Lisbon sailings was not by and large attributable to neglect on the part of the Goa directors. All the ships available to the Goa board had already completed at least one outward passage, and it is clear that many had reached the vice-regal capital in such deplorable condition as to be almost unrepairable. Moreover, the Goa board was to a much larger extent dependent on the good offices of the local crown dockyard authorities, over whom it had no control. A body subordinate to the parent company in Lisbon, it had less freedom of action and ultimately less responsibility than the latter. It was, therefore, able to transfer the blame for many of the faults in naval administration to others, which usually meant the directors in Lisbon or dockyard officials in Goa, with both of whom, it is fair to say, it often squarely belonged.

Instructed by its Lisbon superiors to allow a "small expenditure" only for the repair and fitting out of Indiamen, the Goa board in fact spent quite substantial sums for these purposes—though often apparently still not enough—and in its reports to Lisbon devoted much space to explaining why.[15] The board understandably argued that the ships frequently arrived in India in such poor condition that they required major overhauls before beginning the return voyage to Europe. This claim is convincingly substantiated, at least in the cases of the *Nossa Senhora de Bom Despacho* and *São Gonçalo,* both of which received extensive replacements of masts and spars, and replenishment of stores.[16] On the other hand, the efficient and economical undertaking of such work at Goa was allegedly difficult to arrange. Whereas in the past there had been a number of rival contractors in the city, any one of whom was competent to handle the repairing of large ships, the board pointed out that this was no longer the case, "for all have gone bankrupt and have no money, and there are not more than two men, who are in league with each other."[17] Because these two contractors charged excessively, and also because of a rise in prices of basic materials, dockyard expenses had supposedly doubled.

Account sheets which detail the cost of refitting the three carracks of 1629-1630 in part, though not entirely, confirm this complaint.[18] The alleged monopolist contractors, Vicente Rodrigues and Denis da Costa. together handled work on the *Nossa Senhora de Bom Despacho* and the *São Gonçalo*. Their prices were indeed high, for while both vessels required important replacements of fittings and naval supplies, the cost of such materials was greatly outweighed by the charges for actual contract work. The services of the contractors in each case accounted for almost two-thirds of the total cost of the refit, even though such items as provisions for the return voyage and the replenishment of medical stores were included in the total.

The *Santíssimo Sacramento,* third vessel in the 1629 fleet, required a less elaborate refit and was not contracted out to the same pair as her consorts but to a Hindu named Vitoba Sinai. His service charges were lighter, amounting to just over half the total cost. Thus while the Goa board may have been right in claiming that the local Portuguese contractors were overcharging, its own accounts suggest that there was an Indian contractor available who would undertake some of the company's repair work more cheaply. Nor was Vitoba Sinai's work in any way inferior, since the *Santíssimo Sacramento* was able to sail home in the normal six months, whereas the *São Gonçalo* proved so unseaworthy that she had to seek refuge at Plettenberg Bay in South Africa where she subsequently broke up in a storm, with great loss of life. The *Nossa Senhora de Bom Despacho* eventually lumbered into the Tagus fifteen months after leaving Goa, in an advanced state of decay.[19]

Compounding the dockyard difficulties of the Goa board were the high costs, slow delivery and sometimes unavailability of many essential naval materials in Portuguese India. To improve supply, the board proposed to accumulate its own stocks and lay them in before the annual arrival of the carracks from Europe. The directors suggested that, in order to facilitate this policy, materials particularly difficult or expensive to obtain in Goa, such as anchors and tar, should be shipped out direct from Portugal, and they offered in return to send to Lisbon items made from tropical hardwoods such as rudder shafts, blocks and pumps, which were more cheaply and better made in India.[20] However, these were relatively long-term solutions to the company's supply problems and could not satisfy immediate needs. In practice the company was obliged during its relatively short existence to rely on the royal dockyard at Goa for its naval requirements, despite the delays and high costs involved.

Repeatedly the Goa directors complained that they were forced to seek supplies from crown dockyard officials who deliberately accorded the company's requirements lower priority than they had given the carracks when the crown itself ran the carreira da Índia. They also charged the company higher prices. Typical were the directors' requests to Lisbon in 1630 to order the vedor da fazenda geral and his subordinates to provide them with coir rope "at the same price as they bring it for His Majesty" and also that they be allowed to buy "all the timber and other things needed at the same price as His Majesty buys them."[21] Presumably when the new vedor, José Pinto Pereira, was himself appointed a company director in late 1632, the situation improved. By that time, however, the company's affairs were such that its failure was virtually inevitable.

The Goa directors also complained about the dock and storage facilities the company had been given by Viceroy Linhares. According to its charter the galleys' wharf together with its adjoining warehouses should have been assigned for their use, and these were supposedly handed over to the board shortly after Linhares' arrival in 1629. However, the Goa directors were soon protesting that they had not received all the buildings to which they were entitled. They had been given only one very dilapidated warehouse which was in urgent need of repair, another and better warehouse having been denied them. In the following year the crown ordered Linhares to ensure that the company's rights were fully respected in this regard, but it appears that the second warehouse had still not been handed over when the company finally went into liquidation.[22]

Even if and when the company managed to organize tolerably satisfactory dockyard services and storage facilities, it frequently experienced last-minute difficulties in loading. Such difficulties were attributable not merely to greed, as has sometimes been sweepingly alleged, but to the late arrival at Goa of cargoes from other Asian ports for trans-shipment to the Indiamen, the need to make adequate provision for the sailors' duty-free perquisites, and the difficulties of handling large ships, which had to be loaded downstream near Panjim owing to the relative shallowness of the river Mandovi at Goa itself. In 1630-31 loading of cargo aboard the carracks could not begin until December 20 when the pepper arrived from Cochin. On February 4 and again on February 10 the viceroy reluctantly agreed to postpone the Indiamen's departure for a few days to allow trans-shipment of a recently arrived cargo of cloves from Malacca, but refused subsequent requests for still further delays. The result was that about 200,000 cruzados worth of Cambay textiles were left behind, a loss which Dom

Jorge Mascarenhas later demanded should be made up from Linhares' own pocket.[23]

The viceroy, who took keen personal interest in naval and dockyard affairs at Goa, was repeatedly frustrated and angered by the slow progress of Indiamen being repaired, fitted out and loaded for the return voyage to Lisbon.[24] He was firmly of the belief that it was better for one Indiaman to leave in season (December to early February) than for thirty to sail for home late. Yet in 1630 the *Santíssimo Sacramento,* the *Nossa Senhora de Bom Despacho* and the ill-fated *São Gonçalo* did not leave until March 4—and even then, only on the determined insistence of the viceroy, who overrode the objections of all the ships' officers and went personally to the bar of the Mandovi at 2 A.M. to ensure they set sail on the first favorable wind.[25] Again in 1631 Linhares felt compelled to force the Goa directors to get the carracks off in good time, and they eventually left in mid-February. On this occasion the viceroy's personal interests were involved since he wished to start out on a tour of inspection to Kanara and Malabar, but considered that he could not do so until the Indiamen had left. Later he informed the crown that these ships could have sailed a month earlier had the company not been administering the carreira da Índia, since as merchants the directors were only interested in trade and did not appreciate the importance of departing in season.[26] Finally, in October 1633 Linhares pressured the Goa board to concentrate on the repair and fitting-out of just two rather than all three carracks for that year, since two ships could be adequately prepared by the middle of January whereas the shortage of carpenters and shipwrights, and of cinnamon for the sailors' liberty-chests, meant that three could not be available even by March 15.[27]

While the Goa board admitted that the prices it paid for the refitting of its carracks were high, it was prepared to claim with confidence that the work done was thorough and that the ships emerged very well repaired. This was not, unfortunately, the view of many officers who had to sail these ships back to Europe; nor did the record of the homeward passage during the years of company administration really bear out the board's claims. The *Nossa Senhora de Bom Despacho,* ordered by Viceroy Linhares to leave Goa with the other carracks on March 4, 1630, did so "heavily overloaded, overweighted and leaning to port."[28] The captain-major complained that he had not been present during her refit, having been imprisoned on the viceroy's orders since his arrival from Portugal the previous September. Her pilot and master likewise protested, but the latter's claim that the ship was in no fit condition to leave was ignored by the viceroy. With fourteen feet of water in her hold by the time she had arrived off the Natal coast, the *Nossa Senhora de Bom Despacho* was forced to undergo emergency repairs in False Bay, and again in a second bay east of Cape Agulhas before finally being careened at Luanda.[29] The following year, the officers of the carrack *Santo Inácio de Loiola,* dissatisfied with the repairs their ships had received, "asked for more than was necessary," as the Goa directors defensively put it. The board assured these officers that the ship had received the normal amount of attention in accordance with past custom, but they immediately protested.[30] The *Santo Inácio de Loiola* subsequently ran aground and sank in the Tagus off Oeiras, all but home at the end of her return run.

Understandably, the adequate provision of rations does not seem to have been such a serious problem for ships despatched from Goa for the homeward voyage as it obviously was for those outward-bound. Nevertheless, after Dom Jorge de Almeida's ill-provisioned carracks limped into Goa in 1630, the Goa directors petitioned Viceroy Linhares to have the rations they supplied to the ships for the return voyage inspected by royal officials before departure, in order to ensure that they were adequate.[31] The Goa board's concern over this matter seems to have produced satisfactory results. It is true that in August 1631 the *Santo Inácio de Loiola* had to put into Luanda for fresh provisions, but with ninety-one castaways from the *São Gonçalo* aboard whom she had picked up at sea, there was a good excuse for her doing so.[32] There was, however, much less likelihood of a food shortage on returning carracks than there was on those outward-bound, since ships leaving Goa always had far fewer persons on board and, under the company administration, often carried large stocks of rice in their holds.

Unruliness among the ships' crews was a far more persistent difficulty for the company's board in India.[33] In 1630-31 many men allegedly refused to undertake guard duty on the carracks then loading at Goa, on the grounds that they had bought their places on board from "some officials and other persons" in Lisbon and therefore considered themselves free of further obligations once the carracks had arrived.[34] A more fundamental cause of sailor rebelliousness was the inability of many of the men to make up their customary allowances, partly because of the shortage of cinnamon from Sri Lanka and partly because of insufficient stowage space on board. The loss of their allowances was a serious matter for these sailors, who received few real benefits for their hard, dangerous and widely despised work. Most crew members were in debt to creditors in Lisbon and relied on the proceeds from their liberty-chests to repay what they owed.[35] It was therefore not surprising that many "almost mutinied and refused to go on board" when the carracks were under preparation for the return voyage in late 1630.[36] Even a personal offer from Viceroy Linhares to make available 300 quintals of cinnamon at a reduced price failed to mollify the dissidents who insisted that they would still be unable to pay their debts, and that therefore only one Indiaman instead of two should be sent to Lisbon that year.[37]

There was little either the company or the viceregal authorities could do to coerce the crews in the face of such intransigence. Linhares, though he threatened to hang all who persisted in refusing to embark, also urged the directors to make some sort of provision for those unable to fill their liberty-chests, and declared the handling of the carreira da Índia to be a company, and not a viceregal, responsibility. The Goa directors claimed that it was necessary to treat discontented sailors leniently while they were in India, or they would simply desert. However, when they returned to Portugal they should be treated with severity. The directors went on to name two seamen from the *Bom Jesus de Monte Calvário*, Pero Gonçalves d'Almada and Luís Fernandes, as ringleaders of the malcontents.[38] Whether these two were ever punished is uncertain, but the Goa directors were able to quieten down the sailors and persuade them to sail the carracks back to Portugal in early 1631 only by compensating them with cash from company funds.[39] However, this was a mere palliative to the problem, which persisted and even intensified under the company's administration. In late 1633 Viceroy Linhares advised the company to despatch two instead of three carracks from Goa to Lisbon, partly because this would give the smaller number of sailors required a better chance of filling their liberty-chests. There was special urgency that year to ensure that the sailors received justice, for these particular men had come out to India with Ántonio de Saldanha after the abortive voyages of 1631 and 1632, for which they had had to bear their personal costs. They were therefore "extremely indebted and poor," and it was claimed that if they could not repay what they owed on returning to Portugal, not only would they certainly be harassed and imprisoned by their creditors, but they would be prevented from making further voyages on behalf of Portugal.[40] In the event the third carrack, the giant *Nossa Senhora de Belém,* did remain in India for another year—but still left, undermanned, in 1635 and sank in the Atlantic en route for Lisbon.

It is clear that, if the Goa board of the Portuguese India Company had a rather poor shipping record, it had plenty of excuses. The Goa directors had heavy naval responsibilities, but little power. Faced on the one hand with outworn ships, inadequate and expensive dockyard services, uncooperative officials and discontented crews, they were confronted on the other by the unavoidable fact that to round the Cape successfully Indiamen were bound to leave for Lisbon no later than the recommended sailing season, and by a strong-minded viceroy who tried to insist that they do so. The consequent shipping losses on the homeward passage, although proportionately only slightly heavier than those incurred by returning carracks during the forty years prior to the company's foundation, were more than enough to destroy the hopes of those who had seen the company as the answer to Portugal's trade problems in the Orient. They were also sufficiently damaging to precipitate the company's own ignominious collapse.

9 / Failure and Compromise

The delays and losses at sea which the Portuguese India Company suffered in 1631 and 1632 led inexorably to collapse and liquidation. This consequence of the company's first serious reverses had always seemed probable, given its meagre financial and material resources, the strength of Dutch and English competition, and the deteriorating supply and market conditions for pepper. What made this probability a virtual certainty was the simultaneous failure of the company to achieve satisfactory working relationships with the bureaucracies in Portugal and the State of India. The operations of both the Lisbon board of the company and the Goa board, but particularly the latter, were seriously handicapped by this failure. As far as its India trade was concerned, Portugal was a house divided against itself during the company regime, and suffered accordingly.

When the company was founded in 1628 the crown had clearly instructed that it be given the fullest support and cooperation from officials in India. Promptly following his arrival in Goa in 1629, Viceroy Linhares had issued a decree proclaiming the formation of the local company board, instructing ships' officers and dockyard personnel to cooperate with the board members, and calling on senior officials to give them "all the help and favor they need."[1] In the succeeding years the crown reiterated its instructions that the viceregal government help and protect the company in every way possible. Such support was essential if the Goa board was to function effectively, as the directors themselves were fully aware. "Without his help and favor," they wrote of the viceroy in January 1631, "we shall not be able to effect anything by our own efforts."[2] Initially made up entirely of merchants, of whom several were also the despised New Christians, the board did not possess the prestige to command much respect in an age and society of aristocratic ideals, military preoccupations and ecclesiastical bigotries. Moreover, what the board lacked in standing it scarcely made up in formal authority, of which it enjoyed a very insignificant share, nor in goodwill, since bureaucrats resented it and local businessmen shunned it.

Nevertheless, if the board reports are taken at face value it would seem that in at least the early stages the company did receive from the viceregal government some of the cooperation that it so clearly needed. The viceroy "greatly favors the company," the board reported in 1630, and again the following year, "everything is owing to the Lord Viceroy because he helps all the company's affairs with great zeal." The treasurer-general, Miguel Pinheiro Ravasco, secretary of state Francisco de Sousa Falcão and some dockyard officials were also formally praised by the directors for their helpful attitudes to the company.[3] On closer examination, however, these commendations seem more the result of fear and of a desire to minimize confrontation with the bureaucracy, than of genuine appreciation for services rendered. The Goa directors knew, or soon discovered, that Viceroy Linhares himself did not react kindly to complaints, and they realistically if rather pessimistically believed that future viceroys would be the same. Moreover, the Lisbon board suspected in 1630 that Linhares had begun intercepting confidential company despatches, in part perhaps as a means of influencing the Goa board's communications with Lisbon.[4] In fact, whether thus intimidated or not, the board was clearly reluctant to take a stand against senior officials, and as far as possible endeavored instead to placate them.[5]

In 1630 the Goa directors gave Linhares cargo space on the *Santíssimo Sacramento* without charging freight, urging Lisbon to permit this concession "because if the Lord Viceroy is offended in this we do not know whether he will give as good a reception to the company's affairs as is convenient." In 1631 the directors permitted him to export some ebony to Lisbon on his own account, although this commodity was a company monopoly. Later the same year they found themselves forced to sell the viceregal authorities 577 quintals of pepper for despatch to Europe on the pinnace *Nossa Senhora dos Remédios,* "which we could not resist at all, since the Viceroy wished it of us." Nor did the board press objections when the viceroy himself appointed a new master to the flagship *Santo Inácio de Loiola,* although it thought the right clearly belonged to the company. It also acquiesced even when Linhares placed a guard on each ship since "it seems to us as well to feign agreement, as it is so necessary that the said Lord [Viceroy] be favorable to the company, for without this we will not be able to undertake anything."[6]

Such a relationship meant that in practice crucial decision-making involving company affairs was often exercised by the viceroy and other senior officials rather than by the board. It was the viceroy who determined the conditions under which pepper could be purchased, particularly in Kanara, and how and when it should be brought to Goa.[7] It was the viceroy who could delay or accelerate the departure of Indiamen, who could put pressure on the dockyards to provide proper support to the company, and to whom the directors were obliged to appeal in disputes with the local bureaucrats. Finally, it was the viceroy who possessed de facto power to suspend board members if he saw fit. This relationship might not have been so disadvantageous for the company had Linhares proved its firm supporter, willing and able to do everything possible to promote its work. In fact, the viceroy had little genuine interest in the company and, in so far as he did support it actively, did so out of duty rather than conviction.[8] Linhares' diary and correspondence show that he had an aristocrat's disdain for the merchants who composed the Goa board and regarded them as too committed to selfish business interests to be really trusted with the official India-Portugal trade.[9] The company seemed to him a dubious enterprise which served merely to complicate his extraordinarily difficult task of governing the State of India—and this view eventually became crystallized into the conviction that "the trading company has destroyed India, and if its operations are not stopped, India will be finished."[10]

Despite this scepticism, Viceroy Linhares, who was officially the company's protector—as the crown repeatedly reminded him—always maintained that he meticulously fulfilled his obligations in this capacity.[11] He could have added with justification that his primary viceregal responsibility was to the maintenance of the State of India in general, rather than to the company in particular. His term came at a desperately difficult time, when shrinking Portuguese resources had to be eked out to meet major military crises in Sri Lanka and Mombasa, Malacca and the Persian Gulf, and an extensive famine in northwestern India intensified supply problems. That his government found it genuinely difficult to meet the material needs of the company was, under these circumstances, understandable.

If the viceroy's attitude to the company was initially somewhat lukewarm, that of many subordinate officials proved positively hostile. The Goa directors complained to their Lisbon superiors in December 1630 that they were much hated, and their work deliberately obstructed by the local bureaucracy. Officials allegedly wanted the company's ventures to fail, in the hope of convincing the crown it should never have transferred control of the India trade to mere "businessmen"*—who, moreover, were of doubtful integrity.[12] Not surprisingly, a number of bitter disputes erupted, especially when the company tried to exercise its rights in areas formerly the preserve of local officials whose interests and perquisites were thereby threatened. One such issue concerned the registration of commodities loaded and unloaded from the carracks. Traditionally this had been done by a treasury secretary, but the Goa directors arranged for one of themselves, Fernão Rodrigues de Elvas, to assume responsibility. There were opportunities for personal profit at stake, and João de Sousa de Lacerda, the treasury official concerned, refused to give way, ostensibly on the grounds that to allow a company representative to take over would invite embezzlement.[13] The crown decided firmly in favor of the board on this issue, informing the viceroy that the company should be run by its own personnel, and prohibiting outside interference.[14]

A more serious dispute had arisen in 1629 over the levying of dues at the Goa mint. In accordance with established procedure when the crown itself had operated the pepper monopoly, the Goa board considered that the company's gold brought from Lisbon to buy pepper should be imported tax-free. However, mint officials and the contractor who farmed the right to collect its dues, naturally took the opposite view, and insisted that payments be made. While the viceroy was absent from Goa on a visit to Malabar the company treasurer, Fernão Jorge de Silveira, was accordingly ordered to pay the contractor 300 marks of gold, in lieu of duties not paid on gold already forwarded to Cochin and the Kanara ports. For failing to pay, Fernão Jorge was placed under arrest for almost two months, until Linhares' return to Goa. The dispute was then submitted to the courts but even before a judicial decision had been pronounced, an attempt was made to have sufficient quantities of company pepper forcibly sold to pay the dues claimed.[15]

The crown responded to the company's complaints in this case by reiterating that the Goa directors must not be harassed, but treated with the same respect as government officials.[16] There is, however, ample evidence to demonstrate that the directors were fully justified in

*The term *homens de negócio*—businessmen—had strong New Christian connotations and was clearly used here with anti-semitic intent.

complaining that they were treated with general suspicion and hostility in Goa. In 1631 the câmara itself, which particularly feared possible company participation in the Macao, Malacca and Manila trades, wrote to the crown arguing that the company had proved of little value to either Portugal or the State of India. It also claimed that the company's ships had reinforced Goa with less than 400 men of military age in two years of operations, and that the sailors on the last voyage from Lisbon had complained that rations were so short they would all have died if the ships had remained at sea much longer.[17] These shortcomings had arisen, the Goa Câmara argued, "because the people of the Hebrew Nation [that is, the New Christian directors of the company], being strangers and suspect persons, seek nothing but their own interests and private ends."[18]

To its credit the local company board did from time to time suggest measures which it believed might help to combat more effectively the indifference, lack of cooperation and downright hostility that confronted it in Goa. In 1631 it asked for an official written copy of the company's privileges. It also suggested that all communications between itself and the viceroy and other officials be presented in writing, thus reducing the informal pressures to which the directors felt vulnerable. Another request was for the nomination of a person with whom the directors might lodge confidential protests without the knowledge of the viceroy. They especially emphasized the need for secrecy in implementing this last suggestion, fearing the viceroy's reaction "because if the news should come to him it will be our total destruction."[19]

The most far-reaching of the changes suggested by the board was that its own membership be reduced from five to three, and that one of the three should be the treasurer-general, the highest civilian official in the State of India after the viceroy, who was also responsible for the dockyard. Under this arrangement the treasurer-general would supervise the repair of the carracks and all other needs for which the company depended on dockyard officials. If the ships were his particular responsibility he could be expected to take special care to provide for them. The board added that as long as the treasurer-general was not directly involved in running the company's affairs he was likely to give even minor needs of the crown in India priority over the most vital requirements of the company, but that this attitude would change if he joined the board.[20] Furthermore, since the sailors showed much more respect for the treasurer-general than for the directors, his association with the company's board would improve control over the crews. Finally, his appointment would also have enabled some of the existing directors at Goa to resign, which all were apparently anxious to do.[21]

The crown, following the advice of Dom Jorge Mascarenhas and the Lisbon directors, declined to give the Goa directors the right to retire, but it did in 1632 order the new treasurer-general, José Pinto Pereira, to join the local board. His appointment would undoubtedly have improved the company's standing and effectiveness in India, and have brought the Goa board, as Pereira himself expressed it, greater credit and authority. However, this particular reform, postdating as it did the crucial sailings of 1630-1632, was enacted too late to affect materially the company's declining fortunes.[22]

In certain respects the Lisbon directors were in a much stronger position than their Goa counterparts to combat obstruction and hostility from the company's detractors. This was principally because the Lisbon board's powers and privileges relative to the local authorities were much greater than those of the Goa board. Moreover, Dom Jorge Mascarenhas was an experienced administrator, with the prestige of an aristocrat and personal access to the court in Madrid. Despite these advantages the Lisbon board faced considerable hostility and found that its independence was in reality quite restricted.

According to its charter, the company was subordinated directly and solely to the council of trade. However, it appears that the Duke of Villahermosa, president of the Council of Portugal and one of the inner circle of advisers at court, was the de facto supervising authority over the board. It was Villahermosa who, in the name of the crown, had signed the company's charter in 1628. From at least April 1632 the Lisbon board was formally required to send all its *consultas* to him, and there seems little doubt that in the normal course of events Villahermosa was responsible for the crown's instructions to the company. As president of the Council of Portugal in the mid-1620's, Villahermosa had opposed the idea of forming a company and disputed somewhat acrimoniously with Dom Jorge Mascarenhas on the issue.[23] There is no evidence that this conflict seriously affected the duke's relations with the company board in the 1630's, but some tension between himself and Dom Jorge probably continued. The board received instructions from the crown on a great many matters both important and trivial, ranging from the policies to be adopted for the purchase of pepper to the provision of berths for particular individuals aboard the Indiamen. Sometimes these orders were unrealistic, and involved the company in lengthy explanations as to why they could not be fulfilled. In 1632 the board was told to send four carracks to Goa, but it could only afford to send three. For the following year it was ordered to despatch five, but later this demand had to be reduced to three carracks and a galleon, and finally to just the three carracks.[24]

Dom Jorge Mascarenhas and his fellow directors were also obliged to work closely with the bureaucracy in Lisbon. The board submitted consultas to the Governors of Portugal, and when appropriate these were considered by the local Council of State and Treasury Council. The cooperation of the Lisbon bureaucracy was in fact essential for the board in order to carry out some of its most important tasks, such as the recruitment of soldiers, the provision of emergency shipping when needed, and the fulfillment of sailing dates. In many of these matters, especially the proper management of shipping, the Lisbon government had a vital and legitimate interest since it was primarily on company Indiamen that military reinforcements, cash aids and official despatches were forwarded to Goa. In view of these overlapping functions and interests it was hardly surprising that friction should from time to time occur between Dom Jorge's board and sectors of the Portuguese bureaucracy. In late 1630, in response to such friction, the crown categorically forbade interference in the company's business by government officials, and reminded the Governors of Portugal and the Treasury Council that they had no competence over the affairs of the company, but only the right to consult the crown on any complaint they might have against it.[25]

Mutual suspicion characterized the relationship between directors and local bureaucracy as, for example, in the course of the lengthy efforts of the board and the Treasury Council jointly to draw up an itemized list of the value of the crown ships and naval stores handed over to the company at the beginning of its operations. As early as October 11, 1628, board and treasury representatives, the director of the Armazém, and expert advisers held a preliminary meeting, and work on the valuations began soon afterwards. Discussions continued at irregular intervals in the months and years that followed, but the final inventory was apparently not submitted until April 15, 1633, three days after the company's liquidation.[26] Both the Treasury Council and the board blamed each other for the delays that had occurred, and the treasury representative, João Sanches Baiana, accused the company of deliberately trying to undervalue many of the items concerned.

Complicating these troubles for the company was the development of personal animosity between Viceroy Linhares and Dom Jorge Mascarenhas, which came into the open following the naval inquiry held at Goa in 1630. Evidence was given to the court that Dom Jorge and his sons had been bribed by officers and seamen recruited for the outward-bound Indiamen that year, an allegation Dom Jorge denied and resented.[27] As chairman of the Lisbon board he was able to counterattack by complaining to Madrid that Linhares had refused to allow the homeward-bound carracks of 1631 to delay a few days to pick up a late cargo of cloves and 200,000 cruzados worth of cloth. The viceroy should be forced to compensate the company from his personal estate for the losses it had thereby incurred.[28]

Linhares' response was to ask the crown to give more credence to the consultas of the Goa board than to those of the Lisbon board, and to complain that a director on the latter—unnamed, but obviously Dom Jorge—was hostile to him.[29] Again, in the 1632 despatches, he protested that Dom Jorge Mascarenhas should not be consulted on matters that concerned the Linhares family. The viceroy described Dom Jorge as disaffected towards him and accused him of being a liar and a writer of scurrilities. Subsequently, when yet another inquiry was held into alleged maladministration of the India carracks, Linhares wrote cynically to Dom António de Ataide, count of Castro d'Aire and the then governor of Portugal, that its only result would be to incense Dom Jorge, who would then write more diatribes against him.[30] Animosity between the two officials most responsible for the company's well-being in Goa and Lisbon was hardly likely to ease the company's difficult task, and may help explain why Viceroy Linhares' early suspicions developed into declared hostility against the enterprise by 1632.

Dom Jorge Mascarenhas was on the whole in a strong enough position to defend himself against the criticisms and attacks of enemies, but other members of the Lisbon board, particularly the New Christians, were more vulnerable. One of them, Francisco Dias Mendes de Brito, was arrested and imprisoned sometime before the end of September 1631 by order of the Duke of Maqueda. The reasons for this are unclear, but it appears that the crown intervened curtly on the prisoner's behalf, reiterating that the directors were under royal protection and were exempt from the jurisdiction of other tribunals and officials in all matters concerning the company.[31]

Less easily countermanded was the arrest of another board member, Diogo Rodrigues de Lisboa, ordered by the Lisbon Inquisition on January 11, 1632, on a charge of Judaizing. The evidence came from the confessions of two other Inquisition victims, also New Christians and alleged Judaizers. João Duarte, a Lisbon silk merchant arrested in 1629, had given sworn testimony that, when alone with Diogo Rodrigues de Lisboa in his office on a business matter, the latter had told him in confidence that he believed in the Law of Moses. The second informant, Diogo Mendes de Brito, a businessman and cousin of fellow company director Francisco Dias Mendes de Brito, confessed in September 1631 that three years previously Diogo Rodriques de Lisboa had admitted to him that he secretly trusted in the Jewish faith for his salvation.[32]

As an exceedingly wealthy New Christian entrepreneur of relatively low social status, Diogo Rodrigues de Lisboa was just the kind of person likely to attract Inquisition scrutiny. On the other hand, the fact that he was arraigned on such flimsy evidence suggests a more particular reason for his arrest—and the circumstances of the case further indicate that the probable reason was his selection as a company director. In his statement to the Inquisition Diogo Rodrigues himself claimed that as a director he had many enemies, and that they never hesitated to make scandalous accusations against him and his handling of company affairs. The Holy Office had a vested interest in pursuing any such charges, which would tend to reinforce the Inquisitors' own objections to the terms under which the company had been founded. Unlike the charter of the Brazil Company of 1649, which exempted New Christian investment capital from confiscation in heresy cases, the India Company's charter stopped short of this concession—but it granted such exemption for all other types of offence.[33] Moreover, where a New Christian investor was found to be a heretic, his capital was to be confiscated not to the Inquisition, but to the company. Under these circumstances, the conviction of a company director for Judaizing would have been to the political advantage of the Inquisition, as it would have underlined the dangers to the faith inherent in any relaxation of the laws against crypto-Jews.

The inconclusive nature of the case against Diogo Rodrigues and his persistent denial of guilt persuaded the Inquisitors to test his sincerity by torture. However, when he steadfastly maintained his innocence, despite the rack, the Inquisitors concluded that a major condemnation was not justifiable on the evidence. The prisoner was therefore merely sentenced publicly to abjure his suspected Judaism at the next *auto da fé,* to accept instruction in the faith for the salvation of his soul, and to remain imprisoned at the Inquisition's pleasure. He was probably released following the *auto da fé* on January 9, 1633, about a year after his arrest.[34]

While the unfortunate Diogo Rodrigues de Lisboa was being held by the Holy Office, government officials in Lisbon allegedly discovered that he had been importing coral on his own account from Livorno with the intention of selling it in India, although coral was supposedly a company monopoly. The exposure of this apparent attempt by a director to line his own pockets at company expense drew strong protests to Madrid from the governors of Portugal, who pointed out that the India Company could "hardly take advantage of its coral monopoly if private individuals are trading in it."[35] They then went on to urge that it was high time that the company published its accounts, despite a provision in its charter which allowed it six years before having to do so.

Criticism of the company's management of its finances, and the fact that it had issued no public statement of account despite having received so large an investment from the crown, was mounting in 1632. One unnamed informant, who apparently had access to official records, told Dom António de Ataide that over a period of some three years the company's capital had dwindled from an original 1,200,000 to 600,000 cruzados. He added that the account books of the Casa da Índia and armazém showed that, far from improving the balance of trade as intended, the company had spent more on the fleets and received less from pepper sales, freight charges, and duties than had the previous administration. Like other opponents of the company, this critic accused it not merely of inefficiency, but of corruption. The Lisbon board, he claimed, was dominated by New Christians "who respect only their own self-interest" and who contracted for their naval stores to other New Christians and to their own relatives "without examining the price." Moreover, the board had allegedly spent an average of 170,000 cruzados on each of its Indiamen, whereas a new carrack ready to sail and fully equipped with artillery, sailors, soldiers, provisions and funds for wages, and with 40,000 reals for pepper money, should have cost only 145,000 cruzados. The discrepancy was made even more glaring by the fact that only three of the seven company carracks despatched for Goa to that date were new, and the remainder therefore should not have cost the full amount. Thus the only real result from the creation of the Portuguese India Company had been a change in the administration of the India fleets from the competent hands of the Treasury Council to the supposedly dishonest ones of the directors of the company.[36]

Another critic, whose submission was passed on to Dom António de Ataide by Miguel de Vasconcelos, likewise queried the efficiency and honesty of the company. According to this complainant the company's purchase procedures at the dockyard were highly irregular and suspect. A company-appointed treasurer at the armazém merely had to declare that a particular amount of money was needed for expenditure on the carracks, and without further verification his requisition was authorized. This was despite the fact that one could allegedly expect relatively little honesty from "businessmen who only pay attention to their own individual profits which are their God and their honor."[37]

These criticisms coincided with mounting difficulties for the company in meeting its expenses, which eventually forced Madrid to reconsider the whole enterprise. On November 13, 1632, the governor of Portugal, Dom António de Ataide, wrote to the crown warning that the company would shortly run out of funds. He was thereupon ordered to explain immediately and in confidence why the affairs of the company were not prospering, and to recommend other ways of sustaining the trade with India. A few days later Dom António's response was to inform the king that a balance sheet of the company's profits and losses was already on its way to Madrid, though he understood that this document was "so confused that it will scarcely be possible to understand from it what Your Majesty wishes to know."[38] Meanwhile, a statement issued on December 10 by the company's treasurer declared that it required an estimated 341,396 cruzados to build and fit out the two new carracks needed for the next fleet. Resources likely to be available for the purpose amounted to only 184,866 cruzados, a shortfall of nearly 50 per cent.[39]

The crown now requested a full, comprehensive statement of the company's financial position, and treasurer Pinto da Fonseca hurriedly produced a somewhat inadequate attempt at this on January 3. His report's conclusions were as pessimistic as those reached in the previous

December, and Dom António de Ataide wrote confidentially to the crown two weeks later, arguing that the company had had its day, had proved a failure, and should be wound up as rapidly as possible.[40] In making this recommendation he briefly summarized the history of the company since its inception and enumerated some of the reasons to which he attributed its ill-success. These particularly included its failure to attract any investment from private enterprise and its alleged wastage of money on excessive and duplicated salaries. He argued that the company had only decreased Portugal's trade with India, had consumed all the capital provided and had incurred a net loss. It was left with only two carracks in operation and, even taking into account the funds despatched to India in the last fleet, had insufficient money available to build the two new vessels it needed. He therefore believed the administration of the India trade should be returned to the Treasury Council which, as proved by past experience, could run affairs more profitably and at less cost.[41]

In substance the crown accepted this advice. On April 12, 1633, it declared that the company was in poor condition, that the attraction of private capital, which had been one of the main reasons for its formation, had failed to materialize, and that the royal treasury was unable to finance the fleet for 1634. The company would therefore be liquidated, and the control of the India trade would once more be entrusted to the Treasury Council.[42] An experiment which optimists had once hoped would revive the fortunes of the State of India and the Goa-Lisbon trade had thus ended in dismal failure. With it died any lingering hopes that Madrid and Lisbon may still have nurtured of restoring the good old days of Iberian monopoly on the maritime trade-route from Asia to Europe.

The futility of Hapsburg Portugal's policy of uncompromising military resistance to the intrusions of European rivals into eastern seas was hard to deny by the mid-1620's. The subsequent failure of the Portuguese India Company after less than five years of precarious existence likewise demonstrated Portugal's inability to overcome her imperial problem in the region through economic modernization.

Meanwhile, Portuguese losses in the State of India, particularly to the Dutch, but also to hostile native rulers, Malabari pirates, and the elements, were reaching awesome proportions. A fairly detailed contemporary estimate for roughly the period of Linhares' viceroyalty

(1629-1636) put a minimum value of 7,586,000 xerafins on these losses. At an average of over one million xerafins a year,[43] this figure approximately equalled the regular annual revenues of the entire State of India in the early 1630's and was equivalent to eleven times the annual pepper money sent to Goa from Lisbon by the India Company. With losses of this magnitude, and no realistic prospect of eliminating their enemies by either economic pressure or military force, the only rational alternative left open to the Portuguese, however unpalatable it may have seemed, was peace through political compromise.

The prerequisite for such a compromise was acceptance by the Portuguese of the principle that European nations other than themselves had the right to trade in Asian seas. An early peace agreement, preferably involving an offensive and defensive alliance with either the Dutch or the English, would then have to be sought and hostilities eventually ended with all the rival powers active in the region.

After the loss of Ormuz to the English and Persians in 1622, a strategy of this kind was openly urged by certain members of the Council of Portugal. One councillor, Doctor Mendo da Mota, argued that Portugal could not defend the State of India against the combined naval forces of the English and Dutch, and the land forces of Persia. It was therefore only prudent that the Portuguese ally themselves with one of their Protestant rivals. Pointing out that "peace with England and war with all the world" had been a favorite dictum of the Duke of Alba, Mota went on to urge an alliance with the English.[44]

It was understandable that an alliance with the English in Asia should then be considered more feasible than one with the Dutch, with whom the Twelve Years' Truce (1609-1621) had just expired, and who were still regarded as rebels by the Iberian Crown. Moreover, there was a tradition of Anglo-Portuguese friendship dating back to the late fourteenth century, as well as more recent precedents in Iberian diplomacy for courting England as a counterweight to Holland. By the Anglo-Spanish treaty of 1604 Philip III permitted English merchants to trade freely with Spain and Portugal while banning the Dutch from all such trade early the following year.[45] These measures were intended to stifle Dutch commerce with the peninsula without at the same time impeding fulfillment of the Iberian powers' own needs for the services of foreign merchants and ships. In the event, the Dutch circumvented these restrictions without much difficulty, but attempts were made by both Olivares and the Duke of Buckingham to renew the Anglo-Spanish alliance in 1623 by marrying Charles I of England to a Spanish princess.

In the 1604 treaty no provision was made for a rapprochement beyond the European theatre, but the negotiations undertaken in 1623 temporarily raised hopes in Goa that peace with the East India Company might soon be arranged and at least one of Portugal's major rivals in Asia neutralized.[46] Within a year abandonment of the marriage scheme had dashed these hopes, and the crown renewed instructions to the viceroy in Goa to expel both the Dutch and the English by force.[47] Linhares therefore began his viceregal term—and established the company—committed to an indefinite continuation of hostilities with all Portugal's European rivals, and the possibility of arranging a peace was only revived with the signing of a new Anglo-Spanish treaty in 1630. Official notification of this treaty, together with copies of the text, finally reached Linhares in August 1633, when António de Saldanha's ships arrived at Goa. Meanwhile, by May 1631 reports of the peace were filtering in from the Jesuits in Surat, who were in contact with the English factory, and in October confirmation was received through private merchants' letters arriving overland from Aleppo.[48]

As with the 1604 treaty, the applicability of the peace to Asian seas was not stated in the 1630 agreement, although both Madrid and London were inclined to urge a hard line on their respective local representatives. In early 1631 the East India Company directors resolved to instruct the Surat factory to continue hostilities, while the Iberian Crown informed Linhares in 1634 that the Council of Portugal had advised that the peace could not be extended to India.[49] That serious moves for a truce in the region were nevertheless commenced was due to the initiative of the local Portuguese and English authorities, whose pursuit of it was despite, rather than because of, instructions from the principals in Europe.

The initial peace feeler seems to have been made in May 1631 by Thomas Rastell, president of the East India Company factory at Surat, through the local Jesuit provincial, Padre António de Andrade.[50] Linhares' first reaction was understandably cautious. He had to avoid exceeding his instructions too openly, and yet could not let slip an obvious opportunity of easing the pressures on the State of India at a critically difficult time. The viceroy therefore decided that Andrade should inform Rastell that if the English wanted peace it would have to be on the basis of friendship with Portugal's friends and war with her enemies, particularly the Dutch. At the same time, he assured the crown that he had no intention of arranging an agreement even on these terms, but merely wished to verify the extent of the English offer for the government's information.[51]

The English response made it clear that the Surat factory, though strongly in favor of a truce, was still hesitant to embark on an offensive alliance against the Dutch. This was ostensibly because it had no authority to do so, but in fact probably more because it felt both reluctant and unprepared to take on what could easily prove a perilous commitment. Linhares, perhaps worried by rumors of a large English fleet being prepared in London to blockade Goa, and certainly as anxious as the factory to avoid renewed hostilities, did not break off negotiations but suggested that both parties write home for more precise instructions.[52]

Meanwhile, as Goa awaited official notification of the 1630 treaty, the rumors that began filtering through from Surat were not encouraging for truce prospects. When Padre Paulo Reimão, Linhares' Jesuit agent and himself a Dutchman, asked the commander of the East India Company's ships freshly arrived from England in 1632 about the terms of the peace in Europe he was informed that, although the English negotiators had wanted India included, the Spanish side had demurred. The Council of Portugal had allegedly opposed such a concession on the grounds that India had been won by the sword, and by the sword she would be defended. Linhares ignored this unrealistic bellicosity, merely remarking how delighted he was that the crown now proposed sending him the sword with which to defend India—and continued to pursue his truce negotiations.[53]

On February 3, 1633, the day following receipt of the discouraging reports from Surat, Linhares wrote to the crown in support of a petition from the Goa Câmara asking for peace with the Dutch and English. Although claiming he wrote reluctantly and only after slow deliberation, it is clear Linhares was glad of the opportunity to raise the issue again in this form. He declared that were he given twenty-four well-equipped and properly manned galleons he could overcome Portugal's European enemies in Asia within three to four years. However, if such provision were not made, and made immediately, only a peace or ten-year truce could preserve the State of India.[54]

Further progress towards peace in Asia now depended on the attitude of Madrid, and the long-delayed arrival of the carracks at Goa in

August, bearing copies of the 1630 treaty, finally provided official evidence as to how far this attitude had changed. After discussing the text at length, the viceroy's council decided that the treaty could be interpreted as applying to India and resolved to act on this assumption, pending further instructions from the crown. It therefore agreed that Father Reimão be instructed to show the text to the English president at Surat, and determine his reaction. The padre was then to inform the president that if he made firm proposals for a local truce, the viceroy would agree in principle, and an English representative could subsequently come to Goa to negotiate directly.[55]

The English quickly accepted, and after a further exchange of letters Linhares issued the necessary safe-conducts.[56] Meanwhile, debate on the possibility of an agreement with the East India Company continued at Goa. At his council meeting on March 29, 1634, Linhares himself, while acknowledging that it was beyond normal viceregal competence to negotiate a peace without express royal orders, claimed it was on this occasion appropriate to do so, given the military weakness of the state and the threat of hostile English alliances with Persia and the Moghuls. The meeting concluded by approving the peace moves and resolving that a truce should definitely be signed.[57]

Ironically, only a few days earlier the crown had written to Linhares clarifying that the Council of Portugal would not agree that the Anglo-Spanish peace of 1630 applied to India. The view in Madrid and Lisbon was that by certain articles of both the 1604 and 1630 treaties the English were prohibited from sailing to India and from trading in any Indian ports, whether Portuguese or native. A request was accordingly being made to Charles I of England that he take the necessary steps to remove his vassals from these forbidden regions. In other words, official policy stubbornly held to the old and by now impracticable attitude that the Asian seas were a Portuguese preserve. However, there was a proviso to this apparently inflexible position which gave Linhares the loophole he needed. The despatch from Madrid advised that if the English at Surat wished for a mutual suspension of hostilities while awaiting instructions from their government, the viceroy could agree to such an arrangement, and even in a discreet manner seek it. This expedient was permissible, it was explained, in view of the straitened circumstances of the State of India. The letter went on to emphasize that such a truce must be on a strictly interim basis pending further reference to the crown, and concluded, quite unrealistically, that it should make possible the recovery of Ormuz.[58]

This letter must have reached Linhares aboard the carracks in October 1634, by which time arrangements for English negotiators to come to Goa were already well advanced. Its delivery meant there was no need for Linhares to hesitate further, and when William Methwold, the new English president, finally arrived with three English ships on January 15, 1635, he was received with sincere and elaborate courtesy, though the viceroy himself was confined to his bed with fever. While Methwold emphasized that he could not go to war with the Dutch without orders from Charles I, "which he expected very shortly," the truce, which was signed on January 20, 1635, provided for an indefinite suspension of Anglo-Portuguese hostilities and an alliance against common enemies.[59] In the event of either crown repudiating the truce, six months' notice would have to be given in India before hostilities could be resumed.

Both English and Portuguese in India saw this agreement as advantageous to their interests, particularly vis-à-vis the Dutch.[60] Three weeks after the signing Linhares was writing optimistically to the crown that the parties had agreed to form two squadrons of six Portuguese and six English warships each, to be commanded respectively by a Portuguese and an Englishman. Joint expeditions would be conducted against the Dutch and the Persians, and plans were being prepared for the reconquests of Ormuz and Pulicat. These would be followed by the conquest of the Dutch strongholds in Batavia, Banda, Amboina, Formosa and Ternate.[61]

Such undertakings were almost certainly beyond the capacity of Anglo-Portuguese forces in the region in the mid-1630's even had they been actually attempted. Nevertheless, the agreement brought some real immediate benefits to the State of India. In May 1635 the viceroy was able to charter an English ship to fetch cannon and copper from Macao. These items were urgently needed at Goa but could not be brought out aboard Portuguese vessels because of Dutch patrols in the Malacca Straits. Later the same year Linhares bought more sorely needed copper from the East India Company in exchange for pepper.[62] The English Macao ship was also able to take muskets and munitions of war to Malacca, and private Goa merchants freighted aboard their cargoes for China. Even the religious at Goa appear to have welcomed the truce. A priest remarked in a private letter home in December how God was demonstrating the English peace to be good for the State of India. He explained that the English were Protestants, but were interested only in trade, did not meddle in religious matters, and did not wish to be questioned about their own beliefs.[63]

After the expulsion of the Spanish Hapsburgs from Portugal in 1640, a permanent treaty of friendship with England became an urgent priority for the new Bragança monarchy, struggling to maintain its newly-won throne in Lisbon. Confirmation of the Linhares-Methwold pact was sought, and was obtained in the more comprehensive and permanent Anglo-Portuguese treaties of 1642, 1654 and 1661. What had begun as a mere truce between local leaders therefore proved to be the start of a lasting peace between Englishmen and Portuguese in Asia and Asian seas. From the Portuguese viewpoint, however, the Linhares-Methwold pact was much less successful as an alliance. What the Portuguese could no longer defend or regain for themselves, the English for the most part were unwilling or unable to defend or recover for them. In the final years of Hapsburg rule in Portugal the Dutch effectively crippled what was left of the pepper and other commodity trades between the State of India and Lisbon, by periodic blockades of Goa. These began in earnest soon after Linhares left for Europe in 1636, and prevented any more Indiamen—except a carrack, a galleon and two smaller ships—from leaving for Lisbon for the rest of the Hapsburg era.

Following his succession to the Portuguese throne in 1640 John IV urgently sought peace with Holland, and an agreement was accordingly negotiated in 1641.[64] However, because the representatives of the Dutch East India Company in Asia refused to accept this treaty as binding, peace was not effectively extended there until 1663. In the interval, local Dutch forces, which had already seized Malacca in 1641, succeeded in destroying virtually all the major Portuguese trading interests in South Asia, completing their conquest of Sri Lanka with the fall of Jaffna and Manar in 1658, and of the Malabar ports with the fall of Cochin in 1663. Since the Portuguese were also expelled from their trading posts on the Kanara coast in the 1650's by the nayak of Ikkeri, by the time a new Dutch-Portuguese treaty became operative in Asia in 1663 they had lost all their established pepper outlets and could no longer participate significantly in a trade which had played so vital a role in their Asian enterprise for a century and a half.

The twilight of Portugal's pepper empire was a long and drawn-out affair. The Portuguese fought tenaciously to maintain the integrity of the State of India. Only when commercial modernization proved as fruitless as direct military action in achieving this purpose did they seriously seek compromise and peace with their rivals, and reluctantly accept the inevitability of a greatly reduced role beyond the Cape of Good Hope.

Appendixes, Notes, Glossary, Bibliography, Index

Appendix 1 / Capital Invested in Portuguese India Company of 1628-1633 (in cruzados)

A. Contributed by Crown	
Ships and naval supplies	
Nossa Senhora de Bom Despacho	19,522
Nossa Senhora do Rosário	18,500
Bom Jesus de Monte Calvário	17,000
São Gonçalo	15,024
São Tomé (derrick)	6,000
Masts, spars and sails for above vessels	4,978
Pine logs	3,507
Batalha	1,500
Miscellaneous naval supplies	119,384
Subtotal	205,415
Munitions	
97 bronze cannon	117,898
Gunpowder	9,381
Muskets and arquebuses	4,289
Cannon shot	3,240
89 gun carriages	1,914
7 iron cannon	1,080
Subtotal	137,802
Pepper	
Cargoes from returning Indiamen in 1629 (*Bom Jesus de Monte Calvário, Nossa Senhora do Rosário* and *Batalha*)	250,164
Dues	
Duties and freight charges on private cargoes from returning Indiamen in 1629 (as above)	121,501
Dockyard Services	
Shipbuilding and repair work at crown yards	261,927
Cash	
Pepper money from abortive sailings of 1628	80,000
Total	1,056,809
B. Contributed by the Portuguese Municipalities	
Lisbon	150,000
Évora	16,620
Castelo Branco	16,569
Coimbra	15,000
Viana do Castelo	9,544
Torres Vedras	8,776
Portalegre	7,987

Beja	7,604
Santarém	7,455
Torre de Moncorvo	6,391
Tavira	6,360
Guimarães	6,262
Guarda	6,118
Tomar	6,088
Elvas	5,928
Campo de Ourique	4,953
Lamego	4,886
Setúbal	4,547
Pinhal	4,285
Avís	4,089
Leiria	3,575
Bragança	3,001
Lagos	2,845
Alcobaça	2,736
Viseu	2,470
Crato	2,200
Miranda	1,475
Porto	600
Aveiro	503
Total	318,867
C. Contributed by Private Investors	
Dom Afonso de Furtado Mendonça, Archbishop of Lisbon	500
Dom António de Almeida, Alcaide-mor of Abrantes	1,000
Total	1,500
D. Contributed by Municipalities in the State of India	
Chaul	3,750
Total from All Sources	1,380,926

Sources: Lynch, ff.166-168v, 169-196, 235-250v.

Appendix 2 / Statistics for Lisbon-Goa Trade

2.1. Pepper Money Exported from Lisbon to Goa, 1611-1633 (in cruzados)

A. *Under Crown Control*	*Amount*
1611	100,000
1612	75,000
1613	45,000
1614	90,000
1615	172,000
1616	40,000
1617	201,000
1618	120,000
1619	80,000
1620	80,000
1621	80,000
1622	142,000
1623-1624	208,000
1625	80,000
1626	106,000
1627	40,000
1628	73,000
Subtotal	1,732,000
Annual average	96,222
B. *Under Company Control*	
1629	120,000
1630	80,000
1631	9,128
1632	47,768
1633	102,598
Subtotal	359,494
Annual average	71,899
Total (1611-1633)	2,091,494
Overall annual average (1611-1633)	90,935

Sources: LMG 13A, ff.161-179; LML 38, ff.358-369v; AHU, Documents Soltos, caixa 9, no. 177; Lynch, ff.62-63v, 250.

2.2. Pepper Shipments from Goa to Lisbon, Showing Geographical Origins, 1612-1634 (in heavy quintals)

Year	Kanara	Cannanore	Cochin	Quilon	Malacca	Unidentified	Total
1612	7,025	483	—	—	2,001	—	9,509
1613	4,237	518	585	—	2,583	79	8,002
1614	4,002	—	—	—	—	2,400	6,402
1615	2,633	—	7,722[a]	—	—[a]	—	10,355
1616	11,741	—	4,100[b]	—[b]	—[b]	—	15,841
1617	3,083	—	1,304[c]	—[c]	—	—	4,387
1618	9,602	—	491	—	1,498	—	11,591
1619	10,078	—	814	—	—	—	10,892
1620	5,916	—	83	—	—	—	5,999
1621	3,266	—	4,320	—	—	—	7,586
1622	5,478	—	344	—	—	—	5,878[d]
1623	2,980	—	6,178	—	—	—	9,158
1624-25	17,577	—	3,796	—	—	—	21,373
1626	2,920	—	3,528	3,522	—	—	9,970
1627	5,150	—	9,680[c]	—[c]	—	—	14,830
1628	2,957	—	2,864	—	—	—	5,821
1629	1,790	—	3,715[c]	—[c]	—	—	5,505
1630	1,877	—	3,321	4,686	—	—	10,884
1631	—	—	—	—	—	—	9,061[e]
1632	—	—	—	—	—	—	565[e]
1633	9,686	—	—	—	—	—	9,686
1634	6,152	—	2,893	—	—	—	9,045
Overall total							202,340
Overall annual average							8,797[f]
Total for period under crown control (1612-1629)							163,099
Annual average (1612-1629)							9,061[g]
Total for period under company control (1630-1634)							39,241
Annual average (1630-1634)							7,735[h]

Sources: LMG 13A, ff.161-179; LML 38, ff.358-369v; AHU, Documentos Soltos, caixa 9, no.177; Lynch, ff.57, 59, 62-63v, 89, 227, 239, 250.

[a] Cochin with Malacca.

[b] Cochin with Quilon and Malacca.

[c] Cochin with Quilon.

[d] Includes 56 quintals of unspecified origin seized from illegal traders.

[e] No breakdown available. The 1632 pepper was shipped on behalf of the crown, not the company.

[f] 10,054 light quintals.

[g] 10,355 light quintals.

[h] 8,840 light quintals.

2.3. Purchase Prices of Pepper at Outlets in Southwest India (in xerafins per heavy quintal)

Year	Kanara	Malabar
1607	15-14	12-11
1608-10	— [a]	—
1611	15½	—
1612	15½	—
1613	17	—
1614	18	—
1615	18½	—
1616	19½	11½
1617	19½	—
1618	18½	—
1619	18	12
1620	19	13½
1621	21-16	—
1622	19½-16½	12½
1623	—	12½
1624	17-15	—
1625	16	—
1626	23-15	—
1627-8	—	—
1629	22-15½	10½
1630-1	—	—
1632	22-?	—
1633	—	14
1634	—	—
1635	—	—
1636	?-17	—

Sources: DUP III, 337; LMG 13A, ff.161-179; LML 31, ff.341 and 38, ff.358-369v; AIG, codex 1162, f.204v; Lynch, ff.211, 214.

Note: Kanara prices have been converted from pagodas per candy, or xerafins per candy.

[a] Dashes indicate data not available.

2.4. Portuguese India Company's Imports to Goa, 1629 and 1630 (in milréis)

Import	Lisbon value	Per cent of value of total cargo	Sale price at Goa	Per cent of value of total sales	Profit	Per cent profit
1629						
Silver	29,481	62.00	50,137	63.00	20,656	70.00
Gold coin	2,884	6.00	4,684	6.00	1,800	62.50
Gold bullion	15,184	32.00	24,575	31.00	9,391	62.00
Total	47,549	100.00	79,396	100.00	31,847	67.00
1630						
Silver	22,695	67.75	34,760	75.50	12,065	53.50
Gold	6,439	19.25	10,101	21.75	3,663	56.25
Subtotal	29,134	87.00	44,861	97.00	15,728	54.00
Coral	4,900	13.00	—[a]	—	—	—
Total	34,124	100.00				

Sources: Lynch, ff.58-58v, 62v, 225-225v.
[a]Dashes indicate data not available.

2.5. Portuguese India Company Exports from Goa (1630-1634)

Year	Commodity	Quantity in light quintals (except where otherwise stated)	Value at Goa in milréis	Percentage of total company exports
1630	Pepper	12,439	44,503	81.00
	Indigo	325½	6,390	11.50
	Cowries	1,638 *páras*	1,379	2.50
	Saltpeter	167¾	1,138	2.00
	Rice	1,500 bales	1,129	2.00
	Ebony	297¾	462	0.75
	Timber for ships' parts	—	295	0.50
	Total		55,296	
1631	Pepper	10,356	33,456	96.00
	Saltpeter	95	550	1.75
	Ebony	601	777	2.25
	Total		34,783	
1632	Ebony	50	—[a]	—[a]
1633	Pepper	11,069	49,300	98.00
	Saltpeter	57	907	2.00
1634	Pepper	10,335	—	—

Sources: Lynch, ff.59v, 62-62v, 219, 226-229v; LML 38, ff.366-367v.
[a]Dashes indicate data not available.

2.6. Pepper Loaded at Goa by the Portuguese India Company, 1630-1633 (in light quintals)

Year	Vessel	Quantity	Total
1630	*Santíssimo Sacramento*	4,621	
	Nossa Senhora de Bom Despacho	4,539	
	São Gonçalo	3,279	
			12,439
1631	*Santo Inácio de Loiola*	4,587	
	Bom Jesus de Monte Calvário	5,769	
			10,356
1632	None		
1633	*São Filipe* *São Francisco de Borja*	11,069	11,069
1634	*Nossa Senhora da Saúde* *Santíssimo Sacramento*	10,335	10,335
Total			44,199
Annual Average	(1630-1633)		8,840

Sources: Lynch, ff.57, 59, 62v, 89, 226v-227, 239; LML 38, f.366v.

Note: Although the India Company despatched no pepper in 1632, the Goa government chartered the pinnace *Nossa Senhora dos Remédios,* shipping aboard a small consignment of 646 quintals of pepper, on the state's account.

2.7. Pepper Landed at Lisbon by the Portuguese India Company, 1628-1632 (in light quintals)

Year	Vessel	Quantity	Total
1628	*Madre de Deus*	1,981	1,981
1629	*Bom Jesus de Monte Calvário*	5,649	
	Nossa Senhora do Rosário	4,263	
	Batalha	1,300	11,212
1630	*Santíssimo Sacramento*	4,528	4,528
1631	*Nossa Senhora de Bom Despacho*	2,291	
	Bom Jesus de Monte Calvário	5,382	
			7,673
1632	*Santo Inácio de Loiola*	2,002	2,002
Total			27,396
Annual Average	(1628-1632)		5,479

Sources: Lynch, ff.56-6v, 57, 59-9v, 62-2v, 166-8, 238, 243v-4.

Note: The cargo of the *Madre de Deus* (1628) was purchased from the crown, and the cargoes of the three 1629 Indiamen formed part of the crown's investment in the company.

2.8. Returns to the Portuguese India Company on the Voyages from Goa to Lisbon of Five Indiamen between 1630 and 1632 (in milréis)

Vessel and commodities carried	Purchase price in India	Sale price at Lisbon[a]	Gross profit[b]	Per cent gross profit
Santíssimo Sacramento (1630)				
Pepper	16,532	40,148	23,616	143
Cowries	558	—	—	—
Ebony	223	—	—	—
Saltpeter	407	407	nil	nil
Rice	414	1,461	1,047	253
Indigo	1,767	4,001	2,234	126
	19,001			
São Gonçalo (1630)				
Pepper	11,731			
Cowries	240			
Saltpeter	461		Total Loss	
Rice	339			
Indigo	2,562			
	15,333			
Nossa Senhora de Bom Despacho (1630-1631)[c]				
Pepper	16,241	—	—	—
Cowries	821	—	—	—
Saltpeter	271	271	nil	nil
Rice	377	—	—	—
Indigo	2,061	4,453	2,392	116
	19,772			
Bom Jesus de Monte Calvário (1631)				
Pepper	18,638	47,179	28,541	153
Salpeter	284	284	nil	nil
Ebony	424	—	—	—
	19,346			
Santo Inácio de Loiola (1631-1632)[d]				
Pepper	14,818	19,219	4,401	30
Saltpeter	266	—	—	—
Ebony	353	—	—	—
	15,437			

Source: Lynch, ff.59v, 62-62v, 226-229v.

[a]Much of the cowries and ebony from all vessels remained unsold in the Casa da Índia when the company was liquidated in 1633.

[b]Freight and duties paid to the company on private cargoes totalled a further 23,212 milréis for the *Santíssimo Sacramento,* 7,829 milréis for the *Nossa Senhora de Bom Despacho,* and 15,410 milréis for the *Bom Jesus de Monte Calvário.*

[c]The return on the *Nossa Senhora de Bom Despachos* pepper sold at Lisbon is not stated in the sources, but since much of her cargo was jettisoned en route, it is likely the figure was low.

[d]All the saltpeter and much of the pepper and ebony aboard the *Santo Inácio de Loiola* was lost when she sank in the Tagus, but 2,002 quintals of pepper were subsequently salvaged.

Appendix 3 / Shipping

3.1. Service of Carracks in the Carreira da Índia, 1580-1640[a]

Name	Period of Service	Length of Service[b]		Return Voyages[c]	Aborted Voyages[d]	Ultimate Fate[e]
		Years	Months			
A. Under Crown Administration						
São Tomé (I)	Apr. 1586-Mar. 1589	2	11	1½	0	Wrecked south of Cape Corrientes[f]
Nossa Senhora de Conceição (I)	early 1587-Aug. 1600	14	6	4½	1	Unknown
Santo António	Mar. 1587-mid 1589	2	3	1	0	Vanished at sea[g]
São Cristovão	Apr. 1588-Sept. 1594	6	5	2½	2	Sank in Indian Ocean[f]
São Bernardo	Apr. 1589-Mar. 1592	2	11	1½	0	Vanished at sea[f]
Madre de Deus (I)	Apr. 1589-Aug. 1592	3	4	1½	0	Captured by English fleet off Corvo[f]
São João	May 1590-Oct. 1600	10	5	3½	1	Broken up
São Felipe (I)	Apr. 1593-mid 1600	7	3	2	0	Vanished at sea[g]
Madre de Deus (II)	early 1594	—	1	0	0	Wrecked near Cape Guardafui[f]
São Simão	Apr. 1595-Mar. 1605	10	4	3½	2	Joined Portuguese squadron at Malacca
Nossa Senhora da Luz (I)	Apr. 1595-early 1596	1	—	½	0	Vanished at sea[g]
Madre de Deus de Guadelupe	Apr. 1595-Jan. 1598	1	9	½	0	Accidently burned at Cochin[f]
Nossa Senhora do Castello	Apr. 1597-mid 1599	2	3	1	0	Wrecked on East African coast[g]

São Martinho	Apr. 1597-Nov. 1600	3	7	2	0	Unknown
Nossa Senhora da Paz	Apr. 1598-July 1603	5	3	2	0	Unknown
São Roque	Apr. 1598-late 1604	6	5	2	1	Probably withdrawn as unseaworthy
São Vallentim	Apr. 1600-June 1602	2	2	1	1	Captured by English off Sesimbra[f]
São Francisco	Apr. 1600-Sept. 1607	7	5	2	0	Destroyed by Dutch at Mozambique[g]
Nossa Senhora da Conceição (II)	Jan. 1601-Oct. 1607	6	9	2½	0	Unknown
São Jacinto	Apr. 1601-1608	7		1½	3	Abandoned at Terceira as unseaworthy
Betancor	Apr. 1603-late 1606	3	6	1½	1	Withdrawn from service as unseaworthy
Palma	Apr. 1604-mid 1608	4	3	1	1	Wrecked on reef near Mozambique[g]
Nossa Senhora dos Mártires	Mar. 1605-Sept. 1606	1	6	½	0	Wrecked at Cachopos[f]
Salvação	Mar. 1605-Sept. 1609	4	6	1	0	Sank near Mombasa[g]
Nossa Senhora da Oliveira	Mar. 1605-Sept. 1608	3	6	1	0	Burned near Goa to prevent capture by Dutch[f]
Penha da França (I)	early 1606-July 1610	4	6	2	0	Unknown
Nossa Senhora de Jesus	early 1606-Aug. 1610	4	7	1½	0	Sank at Bahia, Brazil[f]
Nossa Senhora da Ajuda	Mar.-June 1608		3	0	0	Wrecked on West African coast near Mina[g]
Vencimento do Carmo	Mar. 1608-July 1610	2	4	1	0	Unknown
Nossa Senhora da Piedade	Mar. 1609-Aug. 1612	3	5	2	0	Unknown
Guadelupe	Mar. 1609-Oct. 1614	5	7	1	1	Wrecked on reef near Malindi[g]
São Boaventura	Mar. 1609-Mar. 1615	6	—	1	1	Sank in Indian Ocean after developing heavy leaks[f]

3.1 (cont.)

Name	Period of Service	Length of Service[b]		Return Voyages[c]	Aborted Voyages[d]	Ultimate Fate[e]
		Years	Months			
Nossa Senhora dos Remédios	Mar. 1610-early 1616	6		1½	1	Capsized at Goa[f]
Santa Helena (I)	Mar. 1610-Apr. 1613	3	1	1	1	Probably withdrawn as unseaworthy
Nossa Senhora do Livramento	Mar. 1610	1 day	—	0	0	Sank at bar of Tagus on day of departure[g]
São Filipe (II)	Mar. 1611-Nov. 1615	4	8	2	1	Withdrawn as unseaworthy
Nossa Senhora da Nazaré	Mar. 1612-Sept. 1616	4	6	2	0	Unknown
Nossa Senhora do Monte do Carmo	Mar. 1612-Sept. 1613	1	6	1	0	Probably withdrawn as unseaworthy, after battle with Dutch at St. Helena
Nossa Senhora do Cabo	Mar. 1612-Nov. 1618	6	8	2	1	Unknown
Nossa Senhora da Luz (II)	Apr. 1613-Nov. 1615	2	7	½	1	Wrecked on Faial Island (Azores)[f]
Nossa Senhora da Boa Nova	Apr. 1615-Apr. 1621	6		2	0	Unknown
Jesus	Apr. 1615-Nov. 1619	4	7	2	0	Unknown
São Julião	Mar. 1616-Aug. 1616		5	0	0	Wrecked on one of Comoro Islands after battle with English ships[g]
Vencimento	Mar. 1616-Sept. 1617	1	6	1	0	Unknown
Penha da França (II)	Apr. 1617-Oct. 1621	4	6	2	0	Unknown
Guia	Apr. 1617-mid 1620	3	2	1½	0	Remained in India as too unseaworthy for return voyage
São Carlos	Apr. 1618-July 1622	4	3	1	1	Wrecked on reef near Mozambique after

						battle with Dutch ships[g]
Santo Amaro	Apr. 1618-mid 1620	2	3	1	0	Ran aground at Mombasa[g]
Santa Teresa	Mar. 1619-July 1622	3	4	1	0	Wrecked on reef near Mozambique after battle with Dutch ships[g]
Paraiso	Mar. 1619-Nov. 1622	3	7	1	1	Unknown
Conceição (III)	Mar. 1620-Oct. 1625	5	7	½	2	Abandoned at Terceira as unseaworthy
São José	Apr. 1621-July 1622	1	3	0	1	Destroyed by Dutch in battle off East African coast[g]
São Tomé (II)	Apr. 1621-Sept. 1624	3	5	1	1	Used as a derrick in the Tagus
São João Baptista	Mar.-Oct. 1622	—	7	0	0	Beached on East African coast after battle with Dutch[f]
São Francisco Xavier	Mar. 1623-Oct. 1625	2	7	1	0	Wrecked at bar of Tagus[f]
Santa Isabel	Mar. 1623-Jan. 1624		10	0	0	Sank in storm at Mozambique[g]
Chagas	Mar. 1624-Oct. 1625	1	7	1	0	Unknown
Quietação	Mar. 1624-Oct. 1627	3	7	2	0	Unknown
São Bartolomeu	Apr. 1625-Jan. 1627	1	9	½	0	Sank in Atlantic storm[f]
Santa Helena (II)	Apr. 1625-Jan. 1627	1	9	½	0	Sank in Atlantic storm[f]
Total		234	5	79½	26	

(cont.)

3.1 (cont.)

Name	Period of Service	Length of Service[b] Years	Months	Return Voyages[c]	Aborted Voyages[d]	Ultimate Fate[e]
B. Under Company Administration						
Batalha	Apr. 1626-Jan. 1629	2	9	1	0	Withdrawn from service and sold as hulk
São Gonçalo	Apr. 1626-Mar. (?) 1630	4		1½	1	Wrecked in Plettenberg Bay[f]
Bom Jesus de Monte Calvário	Apr. 1627-Oct. 1631	4	6	2	0	Withdrawn from service and sold as hulk
Nossa Senhora do Rosário	Apr. 1628-Sept. (?) 1632	4	5	1	2	Unknown
Nossa Senhora de Bom Despacho	Apr. 1628-June 1631	3	2	1	1	Withdrawn from service and sold as hulk
Santíssimo Sacramento	Apr. 1629-Oct. 1634	5	6	2	1	Unknown
Santo Inácio de Loiola	Apr. 1630-Apr. 1632	2		1	0	Wrecked at Oeiras and and sold as hulk[f]
Nossa Senhora de Belém	Apr. 1631-Feb. (?) 1635	4		½	1	Run aground on coast of East Africa[f]
Nossa Senhora da Saúde	Mar. 1633-Dec. 1637	4	9	2	0	Broken up
Total		35	1	12	6	

Sources: Lynch, ff.55v, 59v, 60, 243, 245v; Cabreira, *Naufragio;* Faria e Sousa, *Ásia,* VI, 379-380, 501-516; Theal, *Records of South-Eastern Africa,* VI, 411; VIII, 139-186; Duffy, *Shipwreck,* pp. 33-47, 54-56, 59; Axelson, *Portuguese in South-East Africa,* pp. 196-205; sailings lists in TT, codex 319; Évora, codices cxvi/1-39 and cxv/1-21; British Library, Add. MS. 20902; BAC, codex 312A; Falcão, *Livro.*

[a]List A is a sampling of sixty carracks that served on the *carreira* during the forty years preceding the Portuguese India Company's foundation in 1629. It includes 90 per cent of all carracks serving in the period 1600-1627. List B includes carracks that were used by the Portuguese India Company for all or part of their service.

[b]The average length of service for carracks under crown and company administration was 3 years, 11 months. The average length of service for the decade 1618-1627 was 3 years, 3 months.

[c]For the purposes of this table, a voyage from Europe to India and back (usually Lisbon-Goa-Lisbon), or vice-versa, is regarded as a "return voyage," a one-way voyage from Europe to India or vice-versa is regarded as a "half voyage." The average number of return voyages per carrack under crown administration was 1.325; under company administration it was 1.333.

[d]An aborted voyage (*arribada*) occurred when a ship, on either its outward or its homeward voyage, because of contrary weather or other unfavorable conditions, failed to round the Cape of Good Hope. It would then be forced to return to its point of departure, or to some intermediate port such as Mozambique, until the following season.

[e]The ultimate fate of all these ships may be summarized as follows: wrecked under crown administration, 28 (47%), under company administration, 3 (33%); destroyed or captured under crown administration, 5 (8%), under company administration, none; withdrawn from service under crown administration, 11 (18%), under company administration, 4 (44%); fate unknown under crown administration, 16 (27%), under company administration, 2 (22%).

[f]Indicates wreck or destruction by enemy on the homeward voyage to Lisbon. One ship accidently burned at Cochin (*Madre de Deus de Guadelupe,* January 1598) and one capsized at Goa (*Nossa Senhora dos Remédios,* early 1616).

[g]Indicates wreck or destruction by enemy on the outward voyage from Lisbon. One ship was deliberately burned to prevent capture by the Dutch (*Nossa Senhora de Oliveira,* September 1608).

3.2. Movements of Portuguese India Company Ships between Lisbon and Goa, 1629-1635

Vessel	Departure date from Lisbon	Outcome of outward voyage	Departure date from Goa	Outcome of homeward voyage
Santíssimo Sacramento	Apr. 3, 1629	Arrived Goa Oct. 21, 1629	Mar. 4, 1630	Arrived Lisbon, Sept. 27, 1630
Nossa Senhora de Bom Despacho	Apr. 3, 1629	Arrived Goa Oct. 15, 1629	Mar. 4, 1630	Arrived Lisbon probably July 3, 1631
São Gonçalo	Apr. 3, 1629	Arrived Goa Oct. 24, 1629	Mar. 4, 1630	Ran aground and sank on Natal coast
Santo Inácio de Loiola	Apr. 17, 1630	Arrived Goa Sept. 30, 1630	Feb. 13, 1631	Sank in Tagus estuary, Mar. 31, 1632
Bom Jesus de Monte Calvário	Apr. 17, 1630	Arrived Goa Sept. 30, 1630	Feb. 13, 1631	Arrived Lisbon Oct. 19, 1631
Nossa Senhora de Belém	Apr. 18, 1631	Voyage abandoned—returned to Lisbon Sept. 14, 1631	—	—
Nossa Senhora do Rosário	Apr. 19, 1631	Same as above	—	—
Nossa Senhora de Belém	May 20, 1632	Voyage abandoned before clearing the Tagus	—	—
Nossa Senhora do Rosário	May 20, 1632	Voyage abandoned before clearing the Tagus	—	—
Santíssimo Sacramento	May 20, 1632	Same as above	—	—
Galleon *São Francisco de Borja*	June 1, 1632	Arrived Goa Oct. 21, 1632	Feb. 1633	Arrived Lisbon, July 11 1633

Naveta *São Felipe*	June 1, 1632	Arrived Goa Oct. 28, 1632	Feb. 1633	Arrived Lisbon, July 11, 1633
Pinnace *Nossa Senhora de Guia*	June 1, 1632	Arrived Goa Nov. 3, 1632	Remained in India	
Nossa Senhora da Saúde	Mar. 6, 1633	Arrived Goa Aug. 19, 1633	Feb. 13, 1634	Arrived Lisbon, Oct. 21, 1634
Nossa Senhora de Belém	Mar. 6, 1633	Arrived Goa Aug. 19, 1633	Feb. 24, 1635	Wrecked on Natal coast, June 30, 1635
Santíssimo Sacramento	Mar. 6, 1633	Arrived Goa Aug. 19, 1633	Feb. 13, 1634	Arrived Lisbon Oct. 21, 1634

Sources: Sailing lists in TT, codex 319; Évora, codices cxvi/1-39 and cxv/1-21; British Library, Add.MS. 20902; BAC, codex 312A; Falcão, *Livro*.

Note: Except where stated otherwise, all ships are carracks. In 1631 two small pinnaces were sent to India out of season in October, after the carracks had aborted their voyages.

Notes

Abbreviations

AHU	Arquivo Histórico Ultramarino, Lisbon
AIG	Archive of the Indies, Panjim
Ajuda	Biblioteca da Ajuda, Lisbon
APO	*Archivo portuguez oriental,* ed. J. H. da Cunha Rivara
BAC	Biblioteca da Academia das Ciências, Lisbon
BF	*Boletim da Filmoteca Ultramarina Portuguesa,* ed. A. da Silva Rego
BIVG	*Boletim do Instituto Vasco da Gama*
BL	British Library (formerly British Museum)
BN Rio	Biblioteca Nacional, Rio de Janeiro
BNL CP	Biblioteca Nacional, Lisbon, Collecção Pombalina
BNL FG	Biblioteca Nacional, Lisbon, Fundo Géral
DHM	*Documentação para a história dos missões do padroado português do Oriente,* ed. A. da Silva Rego
DRI	*Documentos remettidos da India,* ed. A. de Bulhão Pato
DUP	*Documentação ultramarina portuguesa,* ed. A. da Silva Rego
Évora	Biblioteca Pública e Arquivo Distrital, Évora
LMG	Livro das Monções, Archive of the Indies, Goa
LML	Livro das Monções, Arquivo Nacional da Tôrre do Tombo, Lisbon
LP I, II	António Bocarro, "Livro das plantas de todas as fortalezas, cidades e povoações do Estado da Índia Oriental," in *Arquivo português oriental,* ed. A. B. de Bragança Pereira, bk.4, vol.2, pts. 1 and 2
Lynch	Codex Lynch
TT	Archivo Nacional da Tôrre do Tombo, Lisbon

Chapter 1: The Portuguese in Kanara and Malabar

1. This and the next few paragraphs draw on *The Imperial Gazetteer of India* (new ed., 26 vols., Oxford, 1907-1909), I, 37-47, and O. H. K. Spate and A. T. A. Learmonth, *India and Pakistan. A General and Regional Geography* (London, 1967), ch. 1.

2. For the rise of Ikkeri, see Ajuda, codex 51-vii-12, passim; BNL FG, codex 939, passim; António Bocarro, "Livro das plantas de todas as fortalezas, cidades e povoações do Estado da Índia Oriental," in *Arquivo Português Oriental,* ed. A. B. de Bragança Pereira, bk.4, vol.2, pts. 1 and 2 (hereafter referred to as LP I and II), I, 310; *Boletim da Filmoteca Ultramarina Portuguesa,* ed. A. da Silva Rego (Lisbon, 1954 -), no.5, pp.171-174 (hereafter referred to as BF); *Assentos do conselho do estado 1618-1750,* ed. Panduronga S. S. Pissurlencar (5 vols., Bastorá-Goa, 1953-1957), I, 245 (hereafter referred to as *Assentos*); Pietro della Valle, *The Travels of Pietro della Valle in India* (2 vols., London, 1892), II, 243; *Gazetteer of the Bombay Presidency* (Bombay, 1877-1904), XV, pt.2, pp.121-124; Robert Sewell, *A Forgotten Empire* (London, 1900), p.220; W. H. Moreland, *From Akbar to Aurangzeb* (London, 1923), p.2; H. Heras, "The Expansion Wars of Venkatapa Nayaka of Ikeri," *Proceedings* of the In-

dian Historical Records Commission, 1929, pp.106-124. No coherent account of the history of this princedom has yet been written.

3. Della Valle, *Travels,* II, 202, 245; Peter Mundy, *The Travels of Peter Mundy, in Europe and Asia, 1608-1667* (5 vols. in 6, London, 1907-1936), III, pt.1, p.86; Walter Hamilton, *The East India Gazetteer* (2 vols., London, 1828), II, 1.

4. Évora, codex cxvi/1-18, f.45, and BN Rio, codex 2/2-19, ff.364-364v, where it is stated that Kanara pepper was "discovered" in 1565. In fact, the Portuguese knew at least fifteen years before, and probably earlier, that Kanara produced pepper. See Simão Botelho, "Tombo do Estado da India," in *Subsidios para a historia da India portugueza,* ed. R. J. de Lima Felner (Lisbon, 1868), p.258.

5. LP I, 307-311; *Gazetteer of the Bombay Presidency,* XV, pt.2, pp.305-310.

6. Kanara provided over 50 per cent of Portuguese pepper purchases in ten of the years between 1611 and 1626 (LMG 13A, ff.161-179).

7. LP I, 313-317.

8. Ibid., pp.318-320.

9. Ibid., p.316; AIG, codex 7786, ff.288, 289; BN Rio, codex 2/2-19, ff.89v-90.

10. Simão Botelho, "Tombo," pp.257-258; BF no.1, pp.141-143.

11. Some of these letters survive in the royal correspondence. See, for example, AHU, codex 281, f.36.

12. António Bocarro, *Decada 13 da historia da India* (Lisbon, 1876), ch.8, p.40; ch.18, p.77; ch.43, pp.182-183.

13. BF no.11, pp.58, 73, 111; *Assentos,* I, 20; della Valle, *Travels,* II, 314-315; Paulo da Trindade, *Conquista espiritual do Oriente,* ed. F. Felix Lopes (3 vols., Lisbon, 1962-1967), II, 238-242.

14. Della Valle, *Travels,* II, 253-254, 307.

15. *Documentaçao ultramarina portuguesa,* ed. António da Silva Rego (Lisbon, 1960-), III, 379 (hereafter referred to as DUP).

16. LP I, 311-313; *Assentos,* II, 79-83.

17. For example, there were marital and political links between the rajah of Benguel in Kanara and the kollatiri of Cannanore in Malabar. See *Documentos remettidos da India, ou Livros das monções,* ed. A de Bulhão Pato (5 vols., Lisbon, 1880-1935), V, 237 (hereafter referred to as DRI).

18. T. I. Poonen, *A Survey of the Rise of the Dutch Power in Malabar* (Trichinopoly, 1948), pp.32-42.

19. Jan Huyghen van Linschoten, *The Voyage of John Huyghen van Linschoten to the East Indies* (2 vols., London, 1884-1885), I, 67; LP I, 324-325; Poonen, *Survey,* pp.39, 42.

20. Philippus Baldaeus, "A True and Exact Description of the Most Celebrated East-India Coasts of Malabar and Coromandel," in *A Collection of Voyages and Travels,* comp. A. Churchill and J. Churchill (4 vols., London, 1704), III, 622.

21. Linschoten, *Voyage,* I, 67; François Pyrard de Laval, *The Voyage of François Pyrard of Laval* (2 vols. in 3, London, 1887-1890), I, 448; LP I, 324.

22. DUP III, 306; F. P. Mendes da Luz, ed., *Livro das cidades e fortalezas* (Lisbon, 1960), f.45v.

23. LP I, 323-324.

24. Thomas Herbert, *Some Yeares Travels into Divers Parts of Africa and Asia the Great* (London, 1677), p.136.

25. Pyrard, *Voyage,* I, 371, 402, 404; Baldaeus, "True and Exact Description," p.625.

26. Pyrard, *Voyage,* I, 405; Évora, codex cxv/1-5, f.7; della Valle, *Travels,* II, 362.

27. Pyrard, *Voyage,* I, 405-406; LP I, 327. Francisco da Costa entirely omitted Calicut from his detailed account of the Malabar pepper trade—a sure indication of its insignificance as an official source of supply for the Portuguese. See DUP III, 295-361.

28. Mendes da Luz, *Livro das cidades,* ff.47-47v; Pyrard, *Voyage,* I, 434.

29. LP I, 349.

30. Linschoten, *Voyage,* I, 69-70; Baldaeus, "True and Exact Description," p.632.

31. BNL FG, codex 939, pt.2, f.17.

32. LP I, 339-340, 344-345; Mundy, *Travels,* III, pt.1, p.111.

33. Baldaeus, "True and Exact Description," p.632.

34. BNL FG, codex 939, pt.2, f.17; BN Rio, codex 2/2-19, f.364v; LP I, 351; Mundy, *Travels,* III, pt.1, p.111; V. Magalhães Godinho, *Os descobrimentos e a economia mundial* (2 vols., Lisbon, 1963-1965), II, 63.

35. Mendes da Luz, *Livro das cidades,* ff.47-47v; LP I, 350.

36. Mendes da Luz, *Livro das cidades,* ff.47-47v; LP I, 347-348.

37. Codex Lynch, f.227 (hereafter referred to as Lynch).

38. Évora, codex cxvi/1-18, ff.26v-27.

39. LML 12, ff.52-53; Magalhães Godinho, *Os descobrimentos,* II, 63.

40. LP I, 329.

41. Trindade, *Conquista,* II, 333.

42. LP I, 332-333; Trindade, *Conquista,* II, 329-334, 350-355.

43. Poonen, *Survey,* pp.35-37; P. Shungoonny Menon, *The History of Travancore* (Madras, 1878), pp.114-185.

44. Baldaeus, "True and Exact Description," p.643.

45. LP I, Magalhães Godinho, *Os descobrimentos,* 359.

46. Mendes da Luz, *Livro das cidades,* f.48v; LP I, 360-361; II, 67.

47. Lynch, f.227.

48. For identification of these rajahs, see LP I, 359-360; DUP III, 332; *Regimentos das fortalezas da Índia,* ed. Panduronga Pissurlencar (Bastorá, 1951), pp. 218-219; Poonen, *Survey,* pp.37-38.

Chapter 2: Goa and Portuguese Trade

1. LP I, 20, and II, 14-31. The Damão territories were described as stretching 20 or 25 leagues along the coast, with a width of between 2½ and 8 leagues; the Bassein territories covered a strip of coast over 8 leagues long and between 6 and 7 leagues deep. Both contained many crown villages. In the early seventeenth century, the Portuguese in Sri Lanka, as heirs to King Dharmapala of Kotte, held over 4,640 villages. Mendes da Luz, *Livro das cidades,* f.26v; LP I, 134; Tikiri Abeyasinghe, *Portuguese Rule in Ceylon 1594-1612* (Colombo, 1966), p.125.

2. Linschoten, *Voyage,* I, 176.

3. See Mendes da Luz, *Livro das cidades,* ff.17-18, and LP I, 209-212, 265-266, 303-307.

4. Linschoten, *Voyage,* I, 177-179; Pyrard, *Voyage,* II, pt.1, p.28; della Valle, *Travels,* I, 154; LP I, 261; J. A. de Mandelslo, *Voyages celebres e remarquables* (2 vols., Amsterdam, 1727), col.247; Herbert, *Some Yeares Travels,* p.40.

5. Linschoten, *Voyage,* I, 177; Mendes da Luz, *Livro das cidades,* ff.23, 26v, 40v.

6. *Assentos,* I, 252-253, 303-305.

7. J. N. da Fonseca, *An Historical and Archeological Sketch of the City of Goa*

(Bombay, 1878), p.155; Boies Penrose, *Goa—Rainha do Oriente, Goa—Queen of the East* (Lisbon, 1960), p.55.

8. Jean Mocquet, *Travels and Voyages* (London, 1696), bk.4, p.268. In the late seventeenth century Tours had between 60,000 and 80,000 inhabitants (*La grande encyclopédie,* s.v. "Tours").

9. Pyrard, *Voyage,* II, pt.1, p.35. Dom Álvaro da Costa's claim, with reference to the period 1601-1610, that 40,000 Goan vassals of the Portuguese crown could be called upon for military service, seems optimistic (Évora, codex cxv/1-5, p.13). Bocarro's figure of 22,840 for the early 1630's is more plausible (LP I, 306).

10. Della Valle, *Travels,* I, 157; Pyrard, *Voyage,* II, 39.

11. LP I, 222-223; Mandelslo, *Voyages celebres,* col.255; Pyrard, *Voyage,* II, pt.1, pp.39, 65-67; Linschoten, *Voyage,* I, 264, 275-277. For slaves brought into Lisbon on India carracks in 1600, 1601 and 1602, see Simancas, Sec. Prov. 1578, ff.16-27.

12. Linschoten, *Voyage,* I, 175, 252, 256-257; Pyrard, *Voyage,* II, pt.1, pp.27, 38.

13. C. R. Boxer, *Race Relations in the Portuguese Colonial Empire* (Oxford, 1963), pp.58-59, 65.

14. C. R. Boxer, *Portuguese Society in the Tropics* (Madison and Milwaukee, 1965), pp.26-27.

15. Ibid., pp.28-29; Évora, codex cxv/1-5, p.13; LP I, 211-212, 222, 304, 306; LMG 19A, f.96.

16. Pyrard, *Voyage,* II, pt.1, p.125; Mocquet, *Travels,* f.268v; LP I, 223. Shortly before 1612 Francisco Pais, government accountant at Goa, reported that fewer than 1,500 men were available for military service (BN Rio, codex 2/2-19, f.368).

17. BN Rio, codex 2/2-19, f.368.

18. BNL FG, codex 939, pt.2, f.57v.

19. Ibid., ff.54, 71v, 73-73v, 75v, and 77.

20. For example, see LML 12, f.63; 23, f.163; 37, f.117; 38, f.399v.

21. LML 23, f.3. The prevalence of disease at Goa is well attested—for example by Viceroy Linhares in 1631 (BNL FG, codex 939, pt.2, ff.2v, 21).

22. Quoted in Boxer, *Portuguese Society,* p.31.

23. BN Rio, codex 2/2-19, ff.367v-368. See also LP I, 223, and Pyrard, *Voyage,* II, pt.1, p.34.

24. João Ribeiro, *Ribeiro's History of Ceilão* (Colombo, 1909), pp. 395-396; Évora, codex cv/2-7, ff.71-71v.

25. Viceroy Aveiras to Crown, November 17, 1640 (LML 47, f.68).

26. LMG 14, ff.17v-18, and 19B, f.592v; LML 29, f.180, and 30, f.255. For the complaints of Viceroy Redondo in 1617, see Alfredo Botelho de Sousa, *Subsídios para a história militar marítima da India, 1585-1669* (4 vols., Lisbon, 1930-1956), III, 14; and of Viceroy Aveiras in 1643, ibid., IV, 289. Compare also the remarks of della Valle that "the Religious Orders [in Goa] are numerous, and much more than the city needs . . . so great a number of Religious and Ecclesiastical persons is burdensome to this State and prejudicial to the Militia" (*Travels,* II, 415-416).

27. Pyrard, *Voyage,* II, pt.1, p.57.

28. For the cathedral chapter at Goa, see LP I, 244-245.

29. Pyrard, *Voyage,* II, pt.1, pp.53, 59; LP II, 55, 59-60, 62.

30. See Boxer, *Portuguese Society,* p. 37.

31. Pyrard, *Voyage,* II, pt.1, p.26; della Valle, *Travels,* I, 155.

32. Linschoten, *Voyage,* I, pp.185-192; Pyrard, *Voyage,* II, pt.1, pp.39-64; Penrose, *Goa,* pp.51-81. See also the engravings in the first Dutch edition of Linschoten's *Itinerario* (Amsterdam, 1596), reprinted in the Linschoten-Vereeniging edition (The Hague, 1955-1957).

33. LP I, 236-237, 270, 277; Pyrard, *Voyage,* II, pt.1, p.41.

34. LP I, 279-280; Pyrard, *Voyage,* II, pt.1, p.211.

35. For the carrying capacity of Portuguese Indiamen, see Niels Steensgaard, *Carracks, Caravans and Companies* (Copenhagen, 1973), pp.165-166. For Goa-Lisbon sailings, see TT, codex 319; Évora, codices cxvi/1-39 and cxv/1-21; British Library, Add. MS.20, 902; BAC, codex 312A.

36. LP I, 287-288.

37. C. R. Boxer, "The Portuguese in the East, 1500-1800," in *Portugal and Brazil,* ed. H. V. Livermore (Oxford, 1963), pp.221-222.

38. BN Rio, codex 1-13/2-1, f.lv; LP I, 243, 244, 283-284.

39. Ibid., pp.284-285; Abeyasinghe, *Portuguese Rule,* ch.6; Mundy, *Travels,* III, pt.1, p.71.

40. Mendes da Luz, *Livro das cidades,* ff.39v-40; LP I, 281-282; E. Axelson, *Portuguese in South-East Africa 1600-1700* (Johannesburg, 1960), pp.4-7.

41. Steensgaard, *Carracks,* pp.353-358; C. R. Boxer, "Anglo-Portuguese Rivalry in the Persian Gulf, 1615-1635," in E. Prestage, ed., *Chapters in Anglo-Portuguese Relations* (Watford, 1935), pp.125-126.

42. Mendes da Luz, *Livro das cidades,* f.58v.

43. BN Rio, codex 1-13/2-1, ff.1v, 4-4v; C. R. Boxer, *Fidalgos in the Far East 1550-1770* (London, 1968), p.107.

44. Ibid., p.115.

45. See below ch.6, and C. R. Boxer, *Francisco Vieira de Figueiredo: A Portuguese Merchant-Adventurer in South East Asia, 1624-1667* (The Hague, 1967), pp.51-52.

46. See M. N. Pearson, "Indigenous Dominance in a Colonial Economy. The Goa *rendas,* 1600-1670," *Mare Luso-Indicum,* no.2 (1972), pp.61-73; also, his "Wealth and Power: Indian Groups in the Portuguese Indian Economy," in *South Asia,* no.3 (August 1973), pp.36-44, and *Merchants and Rulers in Gujarat* (Berkeley and Los Angeles, 1976), pp. 39-43, 110.

47. Steensgaard, *Carracks,* pp.83-89, 93, 202-205.

48. Mendes da Luz, *Livro das cidades,* ff.59-60v, 102-102v; LP II, 28.

49. Mandelslo, *Voyages,* col.223; APO V, 924; P. M. Joshi, "The Portuguese on the Deccan (Konkan) Coast: Sixteenth and Seventeenth Centuries," *Journal of Indian History* 46 (1968), 78.

50. LP I, 290.

51. Mandelslo, *Voyages,* col.221; J-B Tavernier, *Travels in India* (2 vols., London and New York, 1889), I, 181-182, 184; J. H. da Cunha Rivara, "O Idalxa 1629-1633," *Chronista de Tissuary,* no.2 (February 1866), p.39; W. Foster, *The English Factories in India* (13 vols., Oxford, 1906-1927), *1624-1629,* p.274.

52. See the articles by Cunha Rivara, "O Idalxa 1629-1633," in *Chronista de Tissuary,* no.2 (February 1866), pp.37-43; no.3 (March 1866), pp.65-74; no.4 (April 1866), pp.89-100; no.5 (May 1866), pp.117-121. For an illuminating account of Bijapur-Goa relations, see Linhares' diary, Ajuda, codex 51-vii-12, ff.98v, 107v, 108, 110-111, 124, 138 and passim, and BNL FG, codex 939, pt.2, passim.

Chapter 3: The Structure of the Pepper Trade

1. The most detailed statements are for 1629-30 and occur in Lynch, ff.203-231v. Much briefer accounts for 1621-1633 may be found in LML 38, ff.358-369v.

2. Duarte Barbosa, *The Book of Duarte Barbosa* (2 vols., London, 1918-1921), II, 55-56, 92; Garcia d'Orta, *Coloquios dos simples, e drogas he cousas medecinais da India* (Lisbon, 1963), ff.171v-177.

3. Pyrard, *Voyage,* II, pt.2, pp.355-356; Linschoten, *Voyage,* II, 72-73; Mundy, *Travels,* III, pt.1, p.79.

4. Francisco da Costa, "Relatório sobre o trato da pimenta," DUP III, 293-361.

5. DUP III, 295, 303, 314, 376, 349-353.

6. Ibid., pp.363-379.

7. The description is Garcia d'Orta's, the translation Markham's. See Garcia d'Orta, *Colloquies on the Simples and Drugs of India,* trans. and ed. Sir Clements Markham (London, 1913), p. 369. For a modern description of pepper, see J. W. Purseglove, *Tropical Crops. Dicotyledons* (2 vols., London, 1968), II, 441-450.

8. Mundy, *Travels,* III, pt.1, p.79.

9. *Imperial Gazetteer,* III, 54-56.

10. Mundy, *Travels,* III, pt.1, p.79.

11. DUP III, 350-351.

12. A. Das Gupta, *Malabar in Asian Trade 1740-1800* (Cambridge, 1967), p.22; Menon, *History of Travancore,* pp.96-97.

13. H. Wigram and L. Moore, *Malabar Law and Custom* (Madras, 1900), pp.151, 157-158; Das Gupta, *Malabar,* p.23. For the Tiyan and Moplahs, see Barbosa, *Book,* II, 59, 75. The Nayars are described by several sixteenth-century Portuguese writers including Barbosa (ibid., pp.38-57) and Fernão Lopes de Castanheda, *História do descobrimento e conquista da Índia* (3rd ed., 4 vols., Coimbra, 1924-1933), I, 35-39.

14. Wigram and Moore, *Malabar Law,* pp.150-151.

15. DUP III, 350.

16. Das Gupta, *Malabar,* p.23.

17. Barbosa, *Book,* II, 56.

18. Francis Buchanan, *A Journey from Madras through the Countries of Mysore, Canara and Malabar* (3 vols., London, 1807), II, 455-456, 464-468, 518-523, 530-533.

19. LML 32, f.106.

20. Das Gupta, *Malabar,* p.22.

21. António da Silva Rego, ed., *Documentação para a história das missões do padroado português do Oriente* (12 vols., Lisbon, 1947-1958), X, 160-161 (hereafter cited as DHM).

22. BF no.12, pp. 453-454.

23. Barbosa, *Book,* II, p.56; Baldaeus, "A True and Exact Description," p.622.

24. DUP III, 350-351.

25. Ibid.; LP I, 361. In 1566 the governor at Goa informed the crown that an estimated 20,000-25,000 quintals of pepper per year was illegally shipped out through the Red Sea, compared with only 10,000-12,000 being despatched on the carracks to Lisbon (DHM X, 157).

26. DHM II, 353-354.

27. DHM II, 175; DUP III, 320-321, 379.

28. Magalhães Godinho, *Os descobrimentos,* II, 60; DUP III, 305.

29. DUP III, 351-352; Magalhães Godinho, *Os descobrimentos,* II, 55-56.

30. Lynch, f.211.

31. DUP III, 315-316, 331; LP I, 338; Tavernier, *Travels,* I, 238.

32. LP I, 361; Simão Botelho, "Tombo," p.35; LP I, 339; Tavernier, *Travels,* I, 238.

33. Della Valle, *Travels,* II, 221.

34. The description that follows is based on Francisco da Costa in DUP III, especially pp.315-316, 322-323, 351-352.

35. BN Rio, codex 2/2-19, f.85v.
36. For inflation and its effects on the pepper industry, see Magalhães Godinho, *Os descobrimentos,* II, 47.
37. DUP III, 313.
38. Ibid., pp.316, 339.
39. LP I, 341.
40. DUP III, 310, 316-318; Lynch, f.213.
41. Lynch, ff.213, 214v-215.
42. Ibid., ff.213, 214v. Rice prices calculated from data in Lynch, ff.214v, 218v. The great famine of 1630-31 is described in Moreland, *From Akbar,* pp. 210-219.
43. Lynch, ff.214v, 213, 215.
44. In 1629, security guards employed at Cochin received 15-20 xerafins each, in the Kanara factories 40 xerafins, and at Quilon 40. The Portuguese left in charge at Onor during the factor's absence received 80 xerafins. Ibid., ff.214v, 215; DUP III, 322.
45. Lynch, f.213v.
46. Ibid., ff.213, 215.
47. Ibid., ff.212, 213v.
48. Magalhães Godinho, *Os descobrimentos,* II, 59.
49. DUP III, 309, 310, 375; Lynch, f.211.
50. DUP III, 324, 325; Lynch, f.212.
51. DUP III, 305, 337.
52. From an anonymous description of the coast of India in DUP II, 145.
53. DUP III, 310; Magalhães Godinho, *Os descobrimentos,* II, 65.
54. Simão Botelho, "Tombo," p.25-26; DUP III, 309-310; *Regimentos,* ed. Pissurlencar, pp.217-219.
55. Diogo do Couto, *Da Asia de Diogo do Couto* (Lisbon, 1777-1788), década 6, bk.8, pp.143-148, 180-195.
56. DUP III, 310, 312.
57. Lynch, ff.213v, 215v; DUP II, 145, and III, 331.
58. Magalhães Godinho, *Os descobrimentos,* II, 66.
59. DUP III, 304-305.
60. Linschoten, *Voyage,* II, 225. See also DUP III, 367-368.
61. DUP III, 367-368; F. P. Mendes da Luz, ed., *Regimento da Caza da India,* (Lisbon, 1951), pp.97-98.
62. *Alguns elementos para a história do comércio da Índia de Portugal existentes na Biblioteca Nacional de Madrid,* ed. J. Gentil da Silva (Lisbon, 1950), pp.41, 64-66.
63. F. P. Mendes da Luz, *O conselho da Índia* (Lisbon, 1952), p.519.
64. Magalhães Godinho, *Os descobrimentos,* II, p.47; DUP III, p.341; F. P. Mendes da Luz, *Regimento da Caza da India,* p.97. The *Santíssimo Sacramento* left Goa on March 4, 1630, with 4,620 quintals of pepper on board, and reached Lisbon on September 23 with 4,528 quintals, a loss of just over 2 per cent (Lynch, f.57).
65. This happened to the *São Gonçalo* in 1630, resulting in the deaths of three sailors. See G. M. Theal, *Records of South-Eastern Africa* (9 vols., Cape Town, 1898-1903), VI, 417.
66. Mendes da Luz, *Regimento da Caza da India,* p.98.
67. See Houghton Library, Harvard, MS Portug. 4794F II, ff.65-65v.
68. For this and the following two paragraphs, see Mendes da Luz, *Regimento da Caza da India,* pp.65-66, 69, 95-96, 98-99, 199-200.

Chapter 4: Crisis in the Early Seventeenth Century

1. AHU, codex 281, f.30v. It has often been suggested that the *Estado da Índia* never paid its way after the lootings and conquests of the early sixteenth century were over. See, for example, J. Lúcio de Azevedo, *Épocas de Portugal económico* (Lisbon, 1947), p.151. This view has recently been challenged by António de Oliveira Marques (*História de Portugal* [Lisbon, 1972], pp.467-468). In any event, viceregal governments were usually able at least to "muddle through" their financial difficulties for most of the sixteenth century, whereas in the late Hapsburg and early Bragança period this was no longer possible.

2. Pyrard, *Voyage,* II, pt.1, p.210.

3. Ajuda, codex 51-vii-12, ff.18, 90v, 91v, 101, 103v, 108v and passim; LML 33, f.265; Botelho de Sousa, *Subsídios,* II, 625; III, 6, 7, 9-14, 72, 231; IV, 8-9, 181, 209.

4. Pedro Álvares de Abreu, "Tratado," Inst. Hist.e Geog. Brasiliero Arquivo, lata 73, doc.23, f.2v.

5. In Mendes da Luz, *O conselho da Índia,* p.246.

6. Évora, codex cv/2-7, f.52; Ajuda, codex 51-vii-30, f.121v.

7. LP I, 280-281, 299.

8. LML 24, f.80.

9. LML 33, f.265.

10. Macao was almost entirely independent financially, its administration largely the responsibility of the local câmara. See C. R. Boxer, *Portuguese Society in the Tropics,* pp. 46-49.

11. BN Rio, codex 2/2-19, f.125v; LMG 13A, f.4.

12. For the rich returns from Ormuz in the mid-sixteenth century, see Simão Botelho, "Tombo," pp.90-92.

13. Inst. Hist. e Geog. Brasiliero Arquivo, lata 73, doc.23, f.2v; Ajuda, codex 51-vii-30, f.118v.

14. LP I, 113-115.

15. BN Rio, codex 2/2-19, f.254.

16. LP I, 299-300.

17. In 1635 Viceroy Linhares expressed himself as dissatisfied with the treasury's share of profits from the Japan trade, claiming that the contractor owed the crown a considerable sum under the terms of his 1629 contract. See C. R. Boxer, *The Great Ship from Amacon. Annals of Macao and the Old Japan Trade, 1555-1640* (Lisbon, 1960), pp.138-139.

18. Ajuda, codex 51-vii-12, ff.2v-4.

19. Ibid., f.90 and passim; BNL FG, codex 939, pt.2, ff.39-39v, and passim.

20. Ajuda, codex 51-vii-12, ff.9v, 51v.

21. BN Rio, codex 2/2-19, ff.179-185; LMG 16A, f.235; BF no.9, p.297; Pyrard, *Voyage,* II, pt.1, p.210.

22. AHU, codex 282, f.15.

23. Évora, codex cxvi/1-18, ff.36-37; BN Rio, codex 2/2-19, ff.361-362v.

24. DUP II, 147; LMG 13A, f.79.

25. Pyrard, *Voyage,* II, pt.1, pp.43, 47; LMG 14, f.28v.

26. BN Rio, codex 2/2-19, f.245v; Mendes da Luz, *O conselho da Índia,* p.246.

27. Évora, codex cxvi/2-3, f.69.

28. Ajuda, codex 51-vii-12, f.66.

29. AHU, Documentos Soltos, caixa 10, no.76.

30. BNL FG, codex 939, pt.2, f.95.

31. Ajuda, codex 51-vii-12, ff.14v, 18v, 20v, 26v, 52; BNL FG, codex 939, pt.2,

ff.2, 71v, 104; Botelho de Sousa, *Subsídios,* I, 36-38; II, 6-15, 265, 518-519; III, 149, 150; IV, 8-9, 473-477.

32. Steensgaard, *Carracks,* pp.67, 83, 85-86, 93.

33. George Winius, *The Fatal History of Portuguese Ceylon* (Cambridge, Mass., 1971), p.96.

34. Botelho de Sousa, *Subsídios,* I, 32; C. R. Boxer, *Portuguese Society in the Tropics,* pp.160-164.

35. LMG 14, f.44v.

36. BN Rio, codex 2/2-19, f.368.

37. Ibid., f.183v.

38. AIG, codex 10397, f.70.

39. Ajuda, codex 51-vii-12, ff.20-20v, 61v, 90; BNL FG, codex 939, pt.2, f.95; Botelho de Sousa, *Subsídios,* IV, 209.

40. AHU, codex 218, ff.79-80, and Documentos Soltos, caixa 10, no.39; BNL FG, codex 939, pt.2, f.105v; Botelho de Sousa, *Subsídios,* II, 267, 625-626; IV, 209.

41. LMG 14, f.24.

42. Magalhães Godinho, *Os descobrimentos,* II, 99.

43. Bocarro, *Decada 13,* pp.363-366; Botelho de Sousa, *Subsídios,* II, 359, 567; Steensgaard, *Carracks,* p.94; F. C. Danvers, *The Portuguese in India* (2 vols., London, 1894) II, 263.

44. Simancas, Estado, legajo 435, f.32.

45. AHU, codex 281, f.340.

46. For the naval encounters of this decade, see Botelho de Sousa, *Subsídios,* II, 5-61, 113-18; Mendes da Luz, *O conselho da Índia,* pp.264-278.

47. Mendes da Luz, *O conselho da Índia,* pp.295-298.

48. For example, see the cautious instructions given in the royal commission for the English East India Company's third voyage (1607-1609), to avoid confrontations with "any the subiecte of the kinge of Spaine, or of any other or confederate ffreinde or Alleys, or any other Nacon or people. . . . " *The First Letter-Book of the East India Company 1600-1619,* ed. George Birdwood and William Foster (London, 1893), p.113.

49. For the Anglo-Portuguese wars in the Persian Gulf, see C. R. Boxer in Prestage, *Chapters in Anglo-Portuguese Relations,* pp.46-129. The importance of the fall of Ormuz is discussed in detail in Steensgaard, *Carracks,* pp.154-414.

50. Madrid, 1627, Ajuda, codex 51-vii-30, ff.112-126.

51. Nuno Alvares Botelho reached Goa in 1624 with a squadron of two carracks and one galleon, which subsequently returned to Lisbon, and five other galleons which remained in Asian waters. Viceroy Linhares left Lisbon for Goa in 1629 with three regular India carracks and six galleons. Two galleons, the *São Estevão* and the *Santiago,* sank en route; the three carracks left again for Lisbon in 1630, and the four remaining galleons were retained for the Estado da Índia.

52. Viceroy Linhares, who judging by his diaries spent an inordinate amount of time and energy on dockyard affairs, succeeded in building only two galleons in his seven years of office.

53. February 4, 1633, Ajuda, codex 49-x-28, ff.365, 365v.

54. Botelho de Sousa, *Subsídios,* IV, 181, 209.

55. Ajuda, codex 49-x-28, f.363v.

56. LML 27, f.66.

57. LMG 19A, f.96.

58. Ibid., where Linhares asserts, "I have reminded Your Majesty in all my letters every year that you should place 4000 men in India at a single blow, and thereafter

provide 1000 each year." See also two earlier letters of December 20, 1632 (BF no.9, p.329), and February 4, 1633 (Ajuda, codex 49-x-28, f.365).

59. Botelho de Sousa, *Subsídios,* III, 6-14, 149; IV, 8-9, 181.

60. Ajuda, codex 51-x-2, ff.72v, 250-253; AHU, Documentos Soltos, caixa 10, no.109.

61. Cited in Linschoten, *Voyage,* I, 199, n.6.

62. AHU, Documentos Soltos, caixa 10, no.103.

63. See comment to this effect by Viceroy Pero da Silva, LML 33, f.261, and 35 f.39.

64. Ajuda, codex 51-x-2, f.203v.

65. LML 33, f.261; 35 f.39; 44 f.67.

66. LMG 19B, f.607; AHU, Documentos Soltos, caixa 10, no.103.

67. December 21, 1630, Ajuda, codex 51-vii-12, f.127.

68. BNL FG, codex 939, pt.2, f.5v; Ajuda, codex 51-vii-12, ff.97v, 98.

69. BNL FG, codex 939, pt.2, ff.46-48.

70. Ibid., and BF no.10, p.394. For earlier and later attempts to organize a terço at Goa, see C. R. Boxer, *The Portuguese Seaborne Empire* (London, 1969), p.301, and Évora, codex cv/2-7, ff.70-73v.

71. December 15, 1635 (LML 35, f.3v).

72. BNL FG, codex 939, pt.2, ff.83-86.

Chapter 5: A Company Conceived

1. Magalhães Godinho, *Os descobrimentos,* II, 86-87.

2. Ibid., p.88. Donald F. Lach, *Asia in the Making of Europe* (Chicago, 1965) I, bk.1, p.109, dates the royal monopoly from about 1504.

3. Magalhães Godinho, *Os descobrimentos,* II, 88-92.

4. Diogo do Couto, *Da Ásia* (12 vols., Lisbon, 1777-1788), bk.10, ch. 6, pp. 571-574; Lúcio de Azevedo, *Épocas,* pp. 136-146; H. Kellenbenz, "Autour de 1600: Le commerce du poivre des Fugger et le marché international du poivre," *Annales: Economies. Sociétés. Civilisations,* vol.11, no. 1 (Jan.-March 1956), esp. pp. 1-5; Lach, *Asia,* I, bk.1, pp. 131-139; Magalhães Godinho, *Os descobrimentos,* II, 92-99.

5. Lúcio de Ázevedo, *Epocas,* pp. 137-141; Lach, *Asia,* I, bk.1, pp. 134, 136, 138; Magalhães Godinho, *Os descobrimentos,* II, 99.

6. Magalhães Godinho, *Os descobrimentos,* II, 99.

7. Tito Augusto de Carvalho, *As companhias portugueses de colonização* (Lisbon, 1902), pp. 19-20; *Dicionário de história de Portugal* (3 vols., Lisbon, 1965-1968), s.v. "Compenhias comerciais."

8. For Shirley's proposal, see DUP II, 264-265, 267; for Gomes Solis', see C. R. de Silva, "The Portuguese East India Company 1628-1633," *Luso-Brazilian Review,* 11 (1974), p.155. Francavila's support is described in Carvalho, *As companhias,* p.23.

9. *Elementos para a história do município de Lisboa,* ed. Eduardo Freire de Oliveira (17 vols., Lisbon, 1887-1898), III, 433-434 (hereafter referred to as *Elementos*).

10. Magalhães Godinho, *Os descobrimentos,* II, 99; De Silva, "The Portuguese East India Company," p. 157.

11. João Baptista Lavanha, *Viage de la Catholica Real Magestad del Rei D Felipe N.S. al Reino de Portugal* (Madrid, 1622); *Elementos,* II, 435; III, 56, 67.

12. John Lynch, *Spain under the Habsburgs* (2 vols., New York, Oxford University Press, 1964, 1969), II, 71-72.

13. Ajuda, codex 49-x-28, f. 363v.
14. De Silva, "The Portuguese East India Company," p. 157.
15. Rafael Ródenas Vilar, "Un gran proyecto anti-holandés en tiempo de Felipe IV," *Hispania,* 22 (1962), 542-558.
16. Ibid.; Roland D. Hussey, *The Caracas Company* (Cambridge, Mass., 1934), pp.8-11.
17. *Elementos,* III, 275-276.
18. For example, apart from Dom Jorge's father's service in Portuguese Asia, his brother was captain of Diu and founded the convent of Descalced Carmelites at Goa; his maternal grandfather had been captain of Chaul; and the husband of an aunt, captain of Sofala (Ajuda, codex 49-xii-38, ff. 1, 2, 12, 16).
19. For this and the next paragraph, ibid.; *Nobreza de Portugal* (3 vols., Lisbon, 1960-1961), II, 23, 24; DUP IV, 496, 508, 536; Gastão de Melo de Matos, *Notiçias do terço da armada real (1618-1707)* (Lisbon, 1932), p. 6.
20. José Justino de Andrade e Silva, ed., *Collecção chronológica da legislação portugueza, [1603-1700]* (10 vols., Lisbon, 1854-[1859]), *1627-1633,* pp.204-207, 310 (hereafter referred to as *Coll. chron.*); Newberry Library, Greenlee MS. 124.
21. António Caetano de Sousa, *História genealógica da casa real portuguesa* (Coimbra, 1946-1951), vol. 11, p. 411; C. R. Boxer, *Salvador de Sá and the Struggle for Brazil and Angola 1602-1686* (London, 1952), p. 160.
22. *Elementos,* p.135.
23. Ibid., pp. 129-131, 139-140, 158.
24. Ibid., pp. 185-186, 240.
25. Ibid., p.246; Francisco Manuel de Melo, *Epanáforas de vária história portuguesa* (Coimbra, 1931), pp.119-209.
26. *Elementos,* III, 309, 314-317; Lynch, f. 242v.
27. *Elementos,* III, 317-318.
28. Ibid., pp.334, 411.
29. In 1628 the Lisbon Câmara agreed to provide 200,000 cruzados in aids for the State of India over a period of six years (ibid., pp.295-306). In August 1630 the câmara undertook to raise 100,000 cruzados for the Pernambuco war.
30. Ibid., pp.173-176, 135 n.2.
31. *Coll. chron., 1620-1627,* pp.137, 410.
32. *Elementos,* III, 174, n.1; *Coll. chron., 1620-1627,* p.156, 167.
33. Lynch, f. 242v.
34. *Elementos,* III, p.314.
35. Lynch, ff. 138-139v.
36. Ibid., ff. 3v., 77, 226, 243.
37. *Elementos,* II, 433-434.
38. Ibid., III, 135, 138.
39. DUP II, pp. 549-550; Magalhães Godinho, *Os descobrimentos,* II, 99; De Silva, "The Portuguese East India Company," p. 157.
40. DUP II, 553-554.
41. *Elementos,* III, 278.
42. Ibid., pp.278-279.
43. Ajuda, codex 51-vi-2, f. 152.
44. Rodenas Vilar, "Un gran proyecto," pp. 555-557, and *La politica europea de España durante la Guerra de Treinte Años (1624-1630)* (Madrid, 1967), pp. 119-131, 133-147; J. H. Elliott, *The Revolt of the Catalans. A Study in the Decline of Spain (1598-1640)* (Cambridge, 1963), pp. 226, 274.

45. Ajuda, codex 46-xiii-30, f. 30.
46. Lynch, f. 243.
47. DUP II, 281-286.
48. BAC, MS verm. 634, para. 59.
49. *Elementos,* III, 138-139.
50. Ajuda, codex 50-v-37, ff. 505v, 507v-508.
51. Ibid., ff. 511v-512.
52. DUP II, 549.
53. Lynch, f. 116v; Ajuda, codex 46-xiii-30, f. 35v.
54. Memo from Ferdinand Kron, January 4, 1628 (Ajuda, codex 51-vii-32, f. 241v).
55. DUP II, 553-554; Ajuda, codex 50-v-37, f. 504v.
56. António Domínguez Ortiz, *Política y hacienda de Felipe IV* (Madrid, 1960), pp. 129-131.
57. Ajuda, codex 50-v-37, f. 504v.
58. C. R. Boxer, "Padre António Vieira S.J., and the Institution of the Brazil Company in 1649," *Hispanic American Historical Review,* 29 (1949), 490.
59. BN Rio, codex Pernambuco 1/2-35, ff. 138-139.
60. The struggle took the form of rival *consultas* to the crown (DUP II, 546-560).
61. Lynch, ff. 40-40v.
62. Ibid., f.2.
63. Ibid., ff. 2v, 166-168v.
64. Quotation from memo of Padre António Vieira to John IV in 1644 (BN Rio, codex 1-13/2-6, f. 107).

Chapter 6: The Company Born

1. Mendes da Luz, *O conselho da Índia,* pp. 81-89, 96, 104.
2. Ibid., pp. 29-39.
3. Ibid., pp. 52-53; Lynch, ff. 29-38v; Mendes da Luz, *Regimento da Caza da India,* pp. 21-22.
4. Mendes da Luz, *O conselho da Índia,* pp. 59-68.
5. Kristof Glamann, *Dutch Asiatic Trade 1620-1740* (Copenhagen and The Hague, 1958), p. 7; C. R. Boxer, *The Dutch Seaborne Empire 1600-1800* (London, 1965), p. 24; K. N. Chaudhuri, *The English East India Company. The Study of an Early Joint-Stock Company 1600-1640* (London, 1965) chs. 1 and 2.
6. Ajuda, codex 50-v-37, ff. 503-513.
7. Ajuda, codex 50-v-37, ff. 506, 506v, 509, 512v.
8. Ibid., ff. 508v-509; Lynch, f. 16.
9. *Elementos,* III, 294-295.
10. Ajuda, codex 46-xiii-30, ff. 30v-31; Lynch, f. 54.
11. DRI II, 67-69, 117-120, 171-172, 327-328; III, 315, 480-481; Bocarro, *Decada 13,* p. 176; *Alguns elementos,* ed. Gentil da Silva, p. 65; Luciano Ribeiro, *Registo da Casa da Índia* (Lisbon, 1954), pp. 384, 388, 520.
12. Ajuda, codex 51-vii-32, f. 235v; codex 46-xiii-30, f. 30v.
13. Ajuda, codex 46-xiii-30, ff. 30v-31; *Grande enciclopédia portuguesa e brasileira,* s.v. "Mata. GENEAL."
14. Ajuda, codex 49-xii-28, f. 188; José Calvet de Magalhães, "Duarte Gomes Solis," *Studia,* 19 (December 1966), p. 148.
15. Dominguez Ortiz, *Política y hacienda,* pp. 130-131.
16. TT, Inquisição de Lisboa, processos N7703 and N4884.

17. In 1604 Solis married Violante Mendes de Brito, a daughter of Heitor the elder and sister of Francisco Dias Mendes de Brito. Calvet de Magalhães, "Duarte Gomes Solis," p. 149; J. Gentil da Silva, *Stratégie des affaires à Lisbonne entre 1595 et 1607. Lettres marchandes des Rodrigues d'Évora et Veiga* (Paris, 1956), p. 25.

18. TT, Inquisição de Lisboa, processo N4774. For the Rodrigues de Lisboa family, see Gentil da Silva, *Stratégie,* pp. 133, 137.

19. BNL FG, caixa 201, no. 115.

20. Ajuda, codex 46-xiii-30, f. 31.

21. Ibid., f. 31. Figueiredo Falcão's book, probably completed in 1612, was published in Lisbon in 1859 under the title of *Livro em que se contém toda a fazenda e real patrimonio dos reinos de Portugal, India e ilhas adjacentes e outras particularidades.*

22. Ajuda, codex 46-xiii-30, f. 31.

23. TT, Inquisição de Lisboa, processo N4474, ff. 30, 41v; Dominguez Ortiz, *Política y hacienda,* p. 129; J. Lúcio de Azevedo, *História dos Christãos Novos portugueses* (Lisbon, 1922) pp. 180-181.

24. Magalhães Godinho, *Os descobrimentos,* II, 95-97.

25. Ajuda, codex 50-v-37, ff. 508-508v, and codex 46-xiii-30, f. 30v; Lynch, f. 16; *Elementos,* III, 294-295.

26. The Brazil Company was financed largely with New Christian capital which was exempted from liability to confiscation by the Inquisition, despite the protests of the Holy Office. See Boxer, "Padre Antonio Vieira," pp. 474-497.

27. Lynch, f.54.

28. Ibid., f.54v.

29. *Coll. chron., 1613-1619,* pp. 347-348.

30. Azevedo, *História dos Christãos Novos,* pp. 186-187; Dominguez Ortiz, *Política y hacienda,* pp. 129-131.

31. *Elementos,* III, 130 n.1.

32. Lynch, f. 3v.

33. Lynch, ff. 15-15v, 29-38v; De Silva, "The Portuguese East India Company," pp. 160, 163, 164.

34. Lynch, ff. 16v-17; Ajuda, codex 50-v-37, ff. 510v, 511.

35. Lynch, ff. 2, 17v, 15-15v, 29-38v.

36. Dom Miguel de Noronha's great grandfather, Dom Afonso de Noronha, had been viceroy of India from 1550 to 1554. This viceroy's nephew, Dom Antão de Noronha, was himself viceroy from 1564 to 1568. Linhares' own father, another Dom Afonso de Noronha, had been appointed viceroy in 1621 but was unable to take up office when his fleet was forced back into Lisbon by Atlantic storms (*Tratado de todos os vice-reis a governadores da India,* comp. A.E.M. Zúquete [Lisbon, 1962], pp. 109-110, 119-120; Évora, codex cxv/1-21, f. 80; TT, cota 319, f. 127).

37. The condition imposed upon Dom Miguel by Dom Fernando was that he marry the latter's niece and heiress, Dona Inácia de Meneses, a condition with which Dom Miguel complied (Ajuda, codex 49-xii-39, ff. 59, 65, 67).

38. The inheritance of the Sá estates was, however, disputed since Dona Felippa willed her properties to the Jesuits (BNL CP, codex 313, f. 163). I am grateful to Mr. Walter Lopes for pointing out that the ensuing litigation is documented at length in TT, Cartório dos Jesuítas.

39. LML 29, f. 35; BF no. 8, p. 82.

40. *Diário do terceiro conde de Linhares, vice-rei da Índia* (Lisbon, 1937-1943), pp. 233-234.

41. Ajuda, codex 51-vii-12, f. 18v, and passim.
42. In the course of his viceroyalty Linhares personally inspected the Goa territories and the settlements and fortresses on the Malabar and Kanara coasts (1631). At various stages he also contemplated a visitation to Portuguese possessions on the northwest coast of India, and personally leading military expeditions to Sri Lanka and Sumatra (ibid., f. 71; BNL FG, codex 939, pt.2, ff. lv-2, 25v).
43. LP I, 255, 292; Manuel de Faria e Sousa, *Ásia portuguesa* (6 vols., Porto, 1947), VI, 437; Ajuda, codex 51-vii-12, f. 57, and passim. The essential structure of Linhares' bridge between Panjim and Ribandar remains today, and is still the main link to Old Goa and the towns and villages farther east.
44. *Elementos,* III, 314 n.3.
45. Ricardo Michaelo Telles, "Brazões de armas nas sepulturas no distrito de Goa," BIVG, 31 (1936), 105; A. C. Germano da Silva Correia, *História da colonização portuguesa na Índia* (Lisbon, 1960), III, 483; Ajuda, codex 49-xii-38, f. 979; C. R. Boxer, "José Pinto Pereira, Vedor da Fazenda Geral da India," *Anais da Academia Portuguesa de História,* VII (1942), 118, n.2.
46. DRI V, 45; AIG, Merces Gerais codex 415, f. 58; LMG 13B, ff. 392-415; AHU, codex 218, ff. 5v, 44; BF no. 5, p. 175; Silva Correia, *História de colonização,* III, 484; Boxer, *Fidalgos,* p. 110.
47. Telles, "Brazões," p. 105; LMG 13B, f. 298; Ajuda, codex 49-xii-38, f. 979; Caetano de Sousa, *História genealógica,* V, 227; José F. Ferreira Martins, *História da Misericórdia de Goa* (Nova Goa, 1910-1914), II, 34.
48. Lynch, f. 118v.
49. AHU, Documentos Soltos, caixa 10, no. 94.
50. LMG 13B, ff. 392v, 396v-397, 398v, 403-403v, 405, 411, 413v; *Diário do terceiro conde de Linhares,* p. 327; AIG, codex 7738, ff. 44-46; BN Rio, codex 2/2-19, f. 97v; Lynch, f. 216; *Assentos,* I, 215.
51. Ajuda, codex 51-vi-21, ff. 148-148v.
52. Lynch, f. 118v.
53. Ibid., f. 122v.
54. AHU, Documentos Soltos, caixa 10, no. 93.
55. Lynch, f. 54v.
56. LMG 13B, ff. 392v, 394v, 396v and passim; AIG, codex 7738, f. 127v.
57. *Assentos,* I, 214 n.1; Lynch, ff. 119, 122v; Ajuda, codex 46-xiii-30, ff. 38, 42v-43.
58. Lynch, f. 122v.
59. LMG 14, ff. 166-167; C. R. Boxer, "On a Portuguese Carrack's Bill of Lading," *Biblos,* 14 (1938), 203.
60. Lynch, f. 121.
61. BF no. 8, p. 209; no. 10, p. 421; LML 30, f. 258; Évora, codex cxvi/2-3, ff. 69-70; De Silva, "The Portuguese East India Company," pp. 168-169.
62. *Assentos,* I, 216.

Chapter 7: The Goa-Lisbon Trade under Company Administration

1. *Elementos,* III, 129; Ajuda, codex 50-v-37, f. 503.
2. LML 27, ff. 521-524v.
3. LML 27, f. 520; Lynch, ff. 136-137; BF no. 7, pp. 572-573.
4. BN Rio, codex 2/2-19, f. 359.
5. BN Rio, codex Pernambuco 1-2-36, f. 247.

6. See especially proposals of António Fialho Ferreira in *Alguns elementos,* ed. Gentil da Silva, pp. 92-97.
7. Ajuda, codex 51-vii-32, ff. 235v-241v.
8. Ibid., f. 234v.
9. Ajuda, codex 50-v-37, ff. 504, 506.
10. Linschoten, *Voyage,* I, 11. Silver reals became dominant in the Goa-Lisbon trade only after about 1550. Previously copper had been the most important single item, and merchandise exported had exceeded specie in value. Magalhães Godinho, *Os descobrimentos,* I, 280-282.
11. Based on data in Lynch, ff. 58-58v, 62v, 225-225v; LML 38, ff. 366-367v; BN Rio, codex 2/2-19, f. 53v. See Appendix II.4.
12. Magalhães Godinho, *Os descobrimentos,* I, 450-451, 456; BN Rio, codex 2/2-19, f. 58.
13. BN Rio, codex 2/2-19, f. 62. It did not follow that gold had come to exceed silver in the total value of official specie and bullion exports to Goa. In 1627, besides the 40,000 cruzados shipped out as pepper money an additional 80,000 cruzados was sent as a grant in aid to the State of India—almost entirely in silver reals (ibid., f. 59).
14. LP I, 279.
15. Foster, *English Factories, 1634-1636,* p. 148.
16. Magalhães Godinho, *Os descobrimentos,* I, 283-285.
17. Lynch, f. 111; Ajuda, codex 46-xiii-30, ff. 36-36v. This loan was probably facilitated by the fact that Manuel de Moraes Supico, a director of the company at Goa, was also president of the local misericórdia.
18. Ajuda, codex 46-xiii-30, f. 36; Lynch, f. 212.
19. Foster, *English Factories, 1634-1636,* p. 121.
20. Lynch, ff. 205-205v; BN Rio, codex 2/2-19, ff. 54, 56v, 65v.
21. Lynch, ff. 211, 212.
22. Lynch, f. 58v; LML 38, f. 367v.
23. Lynch, ff. 64-66, 236.
24. Ibid., ff. 83, 118v; LML 38, ff. 366v-367v.
25. Lynch, f. 250; LML 38, f. 367v.
26. *Alguns elementos,* ed. Gentil da Silva, p. 64; DUP III, 370; Lynch, f. 82v.
27. Lynch, f. 127v.
28. BF no. 10, p. 403.
29. See Appendix 2.1; for the 1580-1584 figure, see Magalhães Godinho, *Os descobrimentos,* II, 96.
30. Lynch, ff. 73, 76, 89-89v, 111v.
31. LML 30, f. 280.
32. Ajuda, codex 46-xiii-30, ff. 44v-45.
33. By comparison the Dutch relied much less heavily on their (mainly Indonesian) pepper, although this commodity was the Dutch East India Company's most important Asian export for much of the seventeenth century, sometimes absorbing over 50 per cent of investment capital (Kristof Glamann, *Dutch-Asiatic Trade, 1620-1740* [The Hague, 1958], p.73).
34. Figueiredo Falcão, *Livro,* pp. 5, 6; DUP III, 364.
35. Magalhães Godinho, *Os descobrimentos,* II, 103-107.
36. Lynch, f. 212v; BF no. 9, pp. 323-324. Apparently no more Indonesian pepper was exported to Europe by the Portuguese after 1617, presumably because of Dutch intervention.
37. LML 31, f. 341.

38. Della Valle, *Travels,* II, 213.

39. BN Rio, codex 2/2-19, f. 89; LML 30, f. 270; BF no. 10, p. 511; *Assentos,* I, 569-571; II, 12-13; *Diário do terceiro conde de Linhares,* pp. 228-229, 236; Foster, *English Factories, 1637-1641,* p. 31.

40. A substantial decline in European pepper prices from about 1620 was also experienced by the Dutch East India Company (Glamann, *Dutch-Asiatic Trade,* pp. 76-80).

41. LML 31, f. 341.

42. Lynch, ff. 59v, 62, 62v.

43. Linschoten, *Voyage,* II, 91.

44. Moreland, *From Akbar,* pp. 108-109; De Silva, "The Portuguese East India Company," p. 188; Pyrard, *Voyage,* II, pt. 1, p. 246.

45. Lynch, ff. 83v., 219; Foster, *English Factories, 1630-1633,* p. 59. See also Appendixes 2.5 and 2.8.

46. Moreland, *From Akbar,* p. 113.

47. Lynch, f. 218v.

48. Ibid., ff. 118, 121-121v.

49. The best description of the Maldivian cowrie trade is by Pyrard, who spent the years 1602-1607 marooned in the archipelago (*Voyage,* I, 236-240).

50. Mendes da Luz, *Livro das cidades,* ff. 97v-98.

51. BN Rio, codex 2/2-19, f. 124v.

52. Lynch, ff. 115v, 216.

53. Linschoten, *Voyage,* I, 33.

54. Lynch, ff. 115, 216v, 227v.

55. Ibid., ff. 59-59v, 62-62v, 126v-127; BF no. 9, pp. 278-279.

56. Moreland, *From Akbar,* p. 119; Magalhães Godinho, *Os descobrimentos,* II, 110. For examples of requests, see AHU, codex 281, ff. 9v, 113, 194.

57. Lynch, ff. 59v, 62-62v, 227v-228.

58. Ibid., ff. 89-89v, 115, 128.

59. Ibid., ff. 227v, 115; Ajuda, codex 46-xiii-30, ff. 43-43v; LML 31, f. 273.

60. Manuel de Faria e Sousa, *Ásia,* VI, 437; Ajuda, codex 51-vii-12, ff. 43, 57, 60, 163 and passim.

61. LML 28, f. 46.

62. Lynch, f. 218.

63. Ibid., ff. 123-123v; LML 30, f. 258.

64. LML 27, f. 523.

65. *The Tragic History of the Sea, 1589-1622,* ed. C. R. Boxer, pp. 3-4.

66. Lynch, f. 217.

67. Ibid., ff. 59v, 62-62v.

68. LMG 13B, ff. 392-405; BN Rio, codex Pernambuco 1-2-36, f. 33. Michael Pearson calculated from the 1630 customs list that almost 4,500,000 yards of cloth were freighted to Portugal on the carracks in 1630 ("Indigenous Dominance," p. 72).

69. LML 33, f. 93.

70. BN Rio, codex Pernambuco 1-2-36, f. 33.

71. Lynch, ff. 73, 224-224v.

72. Ibid., ff. 73v-74; Ajuda, codex 46-xiii-30, ff. 19-19v.

73. Lynch, f. 74; Bernardo Gomes de Brito, *História trágico-marítima,* ed. Gabriel Pereira (12 vols., Lisbon, 1904-1909), XI, 108.

74. Lynch, f. 123v.

75. BF no. 9, p. 280; AHU, Documentos Soltos, caixa 10, Viceroy Linhares to

Crown, Nov. 12, 1633; LML 33, f. 93.

76. See C. R. De Silva, "Trade in Ceylon Cinnamon in the Sixteenth Century," *The Ceylon Journal of Historical and Social Studies,* new ser., vol. 3, no. 2 (1973), p.27.

77. Mocquet, *Travels and Voyages,* bk. 4, p. 269v.

78. Magalhães Godinho, *Os descobrimentos,* II, 107.

79. For the war in Sri Lanka, see P. E. Pieris, *Ceylon and the Portuguese 1505-1658* (Tellipalai, 1920), pp. 209-266; Winius, *Fatal History,* passim.

80. Lynch, ff. 83-83v; Ajuda, codex 46-xiii-30, ff. 28v-29.

81. BF no. 7, p. 498; AHU, codex 476, f. 45.

82. BNL FG, codex 7640, f. 60v; BF no. 6, p. 224.

83. The proposals were made by Francisco Rodrigues da Silveira (late sixteenth century) and António Pereira (1620's). See C. R. Boxer, "Portuguese and Spanish Projects for the Conquest of Southeast Asia 1580-1600," *Journal of Asian History, 3* (1969), 125; Évora, codex cx/2-7, ff. 54-59.

Chapter 8: Company Shipping

1. The Indiamen inherited from the crown were the *Batalha, Nossa Senhora do Rosário, São Gonçalo, Bom Jesus de Monte Calvário* and *Nossa Senhora de Bom Despacho.* Those built by the company were the *Santo Inácio de Loiola, Nossa Senhora de Belém* and *Nossa Senhora de Saúde.*

2. Boxer, *Tragic History,* p. 4; James Duffy, *Shipwreck and Empire. Portuguese Maritime Disasters in a Century of Decline* (Cambridge, Mass., 1955), pp. 51-54.

3. Joseph de Cabreira, *Naufragio da Nao Nossa Senhora de Bethlem* (Lisbon, 1636). The *Nossa Senhora de Belém* sank en route to Lisbon from Goa in 1635.

4. The citizens of Luanda were "amazed" at the size of the *Nossa Senhora de Bom Despacho* (Duffy, *Shipwreck,* p. 56). Excessive size was given as a reason for the wreck of the *São Gonçalo* in 1630 (Faria e Sousa, *Ásia,* VI, 379-380). Judging by the size of its cargo in 1630, the *Santíssimo Sacramento* was larger than either of those ships (Lynch, f. 62).

5. Lynch, ff. 84, 106v, 128v; LML 20, f. 70.

6. Ajuda, codex 46-xiii-30, ff. 31-32.

7. BF no. 8, pp. 81-82; Lynch, f. 205; Ajuda, codex 46-xiii-30, f. 27v.

8. Ajuda, codex 46-xiii-30, ff. 31-31v. The galleons wrecked were the *Santiago* and *São Estevão.*

9. Lynch, ff. 112, 141. The usual duration of the voyage was about six to eight months (Boxer, *Tragic History,* p. 6).

10. What follows is based on the transcript of the inquiry in Lynch, ff. 141-153v. See also Boxer, "The Naval and Colonial Papers of Dom Antonio de Ataide," *Harvard Library Bulletin* 5 (1951), 46-48.

11. Lynch, ff. 59v-60.

12. Faria e Sousa, *Ásia,* VI, 515.

13. Lynch, ff. 160-160v.

14. For the 1632 case, see ibid., ff. 161-163.

15. Ibid., f. 71v.

16. Ibid., ff. 207v-208, 210-210v; Gomes de Brito, *História trágico-marítima,* IX, 107-108, 127-128.

17. Lynch, f. 83.

18. Ibid., ff. 207v-210v.

19. Ibid., f. 209; Gomes de Brito, *História trágico-marítima,* IX, 107-128.

20. Lynch, ff. 72v, 83, 113.
21. Ibid., f. 85; *Coll. chron., 1627-1633,* pp. 201-202.
22. Lynch, ff. 85, 220; LML 31, f. 345; *Assentos,* I, 216.
23. Ajuda, codices 51-vii-12, f. 139, and 46-xiii-30, f. 37v; BNL FG, codex 939, pt.2, f. 1v.
24. See repeated references to naval and dockyard affairs in Linhares' diaries, for example in entries for 25 of the 30 days in November, 1631 (BNL FG, codex 939, pt.2, ff. 90v-104).
25. AHU, Documentos Soltos, caixa 10, no. 112; Ajuda, codex 51-vii-12, f. 3.
26. LML 29, f. 5.
27. AHU, Documentos Soltos, caixa 10, no. 112.
28. Gomes de Brito, *História trágico-marítima,* IX, p. 107.
29. Axelson, *Portuguese in South East Africa,* p. 201.
30. Lynch, f. 111.
31. Ibid., f. 143; Ajuda, codex 51-vii-12, f. 109v.
32. TT, codex 319, f. 136v.
33. Lynch, ff. 73v, 112; BAC, codex 312A, f. 135; British Library, Add. MS. 20902, f. 163.
34. Lynch, ff. 90-90v.
35. Ajuda, codex 51-vii-12, f. 130; AHU, Documentos Soltos, caixa 10, no. 112.
36. Lynch, f. 117.
37. Ajuda, codex 51-vii-12, f. 130.
38. Ibid.; Lynch, f. 112.
39. Lynch, f. 78.
40. AHU, Documentos Soltos, caixa 10, no. 112.

Chapter 9: Failure and Compromise

1. *Assentos,* I, 216-217.
2. Lynch, f. 116.
3. Ibid., ff. 73, 114, 77.
4. Ajuda, codex 46-xiii-30, f. 17v.
5. Lynch, ff. 119-119v.
6. Ibid., ff. 73-73v, 132, 114.
7. For example, the company complained that Linhares had agreed with the nayak of Ikkeri in 1630 that the Portuguese would pay over 27 pagodas per bahar for Kanara pepper when the real market price was only 14 or 15 pagodas. Thus the company claimed it lost "more than 40,000 xerafins." Ajuda, codex 46-xiii-30, ff. 22-24.
8. Linhares seldom mentions the company in the surviving portions of his diary. On December 3, 1632, he wrote to the crown, "I affirm in truth that since taking over this government I have been unaware where the company's directors weigh the pepper because I have entirely fulfilled Your Majesty's order that I know nothing about those people apart from showing them favor when they ask it of me" (LMG 15, f. 69v).
9. See Ajuda, codex 51-vii-12, ff. 127, 129, 129v; LML 29, f. 15.
10. Linhares to Dom António de Ataide, December 6, 1632 (Ajuda, codex 49-x-28, f. 363).
11. AHU, Documentos Soltos, caixa 10, no. 112; Lynch, f. 116.
12. Ajuda, codex 46-xiii-30, ff. 23v, 25v; Lynch, ff. 119-119v, 126v.
13. Ibid., f. 176v; Ajuda, codex 46-xiii-30, f. 25v. The perquisites of this office were valued at about 500 xerafins per ship (Lynch, f. 116).

14. LML 28, f. 22.
15. Lynch, ff. 82, 125v, 131v; BN Rio, codex 2/2-19, f. 55; BF no. 8, p. 81.
16. Lynch, ff. 126-126v; BF no. 9, pp. 280-281.
17. According to the terms of its charter the company was supposed to transport 300 soldiers to Goa in each Indiaman, and to pay for their wages, provisions, arms and passages, from its own resources. On this reckoning, by February 1631 it should have brought out 1,500 men instead of 400.
18. BN Rio, codex Pernambuco 1-2-36, f. 247.
19. Lynch, f. 131, 122.
20. Ibid., f. 81.
21. Ibid., ff. 117v-118, 119.
22. Ajuda, codex 46-xiii-30, ff. 38, 42v-43; AHU, Documentos Soltos, caixa 10, no. 53; BF no. 23, p. 16.
23. For Villahermosa's relationship with the company, see Ajuda, codex 50-v-37, f. 512v; BF no.23, pp.20-21; De Silva, "Portuguese East India Company," p.157.
24. BF no.23, pp. 11-30; Ajuda, codex 46-xiii-30, ff. 44v-45.
25. *Coll. chron., 1627-1633,* pp. 199-200.
26. Lynch, ff. 43v-44, 46-50v, 169-196.
27. Ibid., ff. 88-88v.
28. Ajuda, codex 46-xiii-30, ff. 37-37v.
29. BF no. 8, pp. 78-79.
30. LML 30, f. 164; BF no. 9, p. 320; Ajuda, codex 49-x-28, f. 363v.
31. *Coll. chron., 1627-1633,* p. 226.
32. TT, Inquisição de Lisboa, processo N4474, ff. 3-4v, 5v.
33. Boxer, "Padre Antonio Vieira," pp. 484-485; Ajuda, codex 50-v-37, f.504v.
34. TT, Inquisição de Lisboa, processo N4474, f. 69.
35. Lynch, ff. 124-124v.
36. Ibid., ff. 54-55.
37. Ibid., ff. 16-18. Miguel de Vasconcelos, the supposedly villainous secretary of the last Hapsburg viceroy of Portugal, Margaret of Parma, was assassinated in the uprising against Spanish rule on December 1, 1640.
38. Ibid., ff. 1-1v.
39. Ibid., ff. 23-24.
40. Ibid., ff. 56-60v.
41. Ibid., ff. 1-7.
42. Ibid., f. 7v; *Assentos,* I, 572.
43. See Agostinho de Santa Maria, *Historia da fundação do real convento de Santa Monica* (Lisbon, 1699), pp. 334-343, where data collected by a "zealous official" are cited (p.334).
44. DUP II, 314-315.
45. Júlio Firmino Júdice Biker, *Collecção de tratados e concertos de pazes* (14 vols., Lisbon, 1881-1887), II, 36-38; Mendes da Luz, *O conselho da Índia,* pp. 288-289.
46. *Coll. chron., 1620-1627,* p. 109; LML 19, ff. 45-45v.
47. LML 23, f. 145; Ajuda, codex 51-vii-12, f. 120.
48. *Assentos,* I, 481-483; BNL FG, codex 939, pt.2, ff. 45-45v; LML 29, ff. 145-145v.
49. LML 31, f. 155; R. M. Bharucha, "Anglo-Portuguese Relations (1600-1605)," Ph.D. thesis, University of Bombay, 1946, p. 315.
50. BNL FG, codex 939, pt.2, ff. 45-45v.
51. LML 29, ff. 145-145v.

52. BNL FG, codex 939, pt.2, f. 88. Linhares himself explained the reasons for the English factory's desire for a truce: the English had "neither fortress nor sure haven" in India, being dependent solely on the goodwill of the Moghul authorities at Surat. They were hard-pressed by Dutch competition, were forced to concentrate on the silk trade with Persia where they bought dear and had to sell cheap, and they could not fill their ships through this trade alone. The use of Portuguese ports and the right to trade there would therefore be most useful to them. LML 29, ff. 145-145v.

53. *Assentos,* I, 482 n.1.

54. Ajuda, codex 49-x-28, ff. 365-365v.

55. *Assentos,* I, 481-483.

56. *Diário do terceiro conde de Linhares,* pp. 29-35; *Assentos,* II, 3 n.1.

57. *Assentos,* II, 3-5.

58. LML 31, f. 155.

59. For the negotiations at Goa, see *Diário do terceiro conde de Linhares,* pp. 261-270; Foster, *English Factories, 1630-1633,* pp. xxxv-xxxvii, and *1634-1636,* pp. vii-viii, 21-23, 89-99.

60. Foster, *English Factories, 1634-1636,* p. 8; Biker, *Collecção,* I, 262-270.

61. BF no. 10, pp. 519-521.

62. LML 34, ff. 3, 5-5v, 63v-64v. Beginning in 1643 the Dutch effectively put an end to Portuguese trade in English ships through the Straits of Malacca (Boxer, *Fidalgos,* pp. 113, 114).

63. Padre Manuel Xavier to Manuel Severim de Faria, Dec. 15, 1635 (BNL FG, codex 7640, ff. 93-94).

64. For the 1641 treaty negotiations, see Winius, *Fatal History,* pp. 55-60.

Glossary

ALFANDEGA Custom-house.

ARMAZÉM DA ÍNDIA The dockyard in Lisbon where carracks were built and fitted out for the *carreira da Índia*.

ARRÁTEL (pl. ARRÁTEIS) Portuguese weight equivalent to a pound. In the seventeenth century it normally consisted of 14 onças in Portugal and 16 onças in the State of India.

ARROBA Portuguese weight of 32 arráteis, equivalent to a quarter of a heavy quintal.

BAHAR, BAR A measure of weight commonly used in South and Southeast Asia, which varied according to commodity and region. A bahar of pepper at Cochin weighed 2 heavy quintals 2 arrobas 10 arráteis.

BANYAN, BANIAN, VANIA An Indian trader, especially a Hindu trader from Gujarat.

CÂMARA Municipal council.

CANARIM (pl. CANARINS) Properly a native of Kanara, but used by the Portuguese to denote a native of the Goa territories.

CANDY, CANDIL A measure of weight used in various parts of South Asia, varying according to commodity and region. In Kanara, a candy of pepper was equivalent to 3 heavy quintals 3 arrobas.

CARREIRA DA ÍNDIA The voyage from Portugal to India, and India to Portugal, via the Cape of Good Hope.

CASA DA ÍNDIA India House, the government agency in Lisbon which supervised trade and communications with the overseas empire and collected customs and other charges on imports.

CASA DA MOEDA Mint.

CASADO A married Portuguese settler.

CONSELHA DA ÍNDIA India Council, royal council responsible for Portuguese colonies, 1604-1614.

CONSELHO DA FAZENDA Treasury Council.

CONSULTA A written report drawn up by a state council or other official body, in accordance with an established form. It contained a summary of the views of each voting member of the body, as well as a formal policy recommendation.

COUNTRY TRADE/REGIONAL TRADE Used here to denote seaborne trade within the Asia-East Africa region, and especially to and from Goa.

CRUZADO Portuguese coin, in the seventeenth century usually silver, with a fixed value of 400 réis. It was the equivalent of about four English shillings.

ESTADO DA ÍNDIA State of India, the Portuguese empire east of the Cape of Good Hope.

FANAM Name for a number of South Indian coins of different and fluctuating values. The Cochin fanam was worth 40 réis in the early seventeenth century.

FIDALGO Minor nobleman or gentleman.

GASALHADO Compartment of a ship available to store cargo for a crew-member or other individual.

JUIZ DO PESO Weight inspector for pepper.

LIBERDADE Duty-free cargo allowance permitted each crew-member on returning Indiamen.

MILRÉIS Monetary unit of 1,000 réis.

MISERICÓRDIA The most prestigious lay fraternity in the Portuguese world, with branches in all parts of the empire, providing charitable and other services.

MOPLAH, MAPPILLA Malabar Moslem, descended from immigrant Arab merchants and local women.

NAVETA A small ocean-going ship.

NAYAK, NAIK Indian chief or prince.

NAYAR, NAIR Member of a superior caste in Malabar, composed of rulers, landowners and soldiers.

ONÇA A Portuguese measure of weight equivalent to an ounce.

PAGODA, PAGODE Portuguese name for a variety of gold coins minted in western and southern India, of varying value; often used interchangeably with "pardao."

PAIOL (pl. PAIOES) Ship's hold; cargo compartment.

PARÁ A measure of weight used in various parts of South Asia, usually equivalent to one fourteenth of a candy.

PARDAO, PARDAU The common Portuguese name for the São Tomé and a variety of other coins minted in Western India.

PATACA A common name in Goa and southern India for the São Tomé and its equivalents.

PATTAMAR, PATAMAR A term used mainly in southern India to denote a runner or courier.

PROVEDOR-MOR Director.

QUINTAL The Portuguese hundredweight, basic unit of weight in the India-Portugal trade. In Portugal the light quintal, subdivided into 128 arráteis of only 14 onças each, was used, but in the State of India the heavy quintal consisting of 128 arráteis of 16 onças each.

REAL (pl. RÉIS) The smallest Portuguese monetary unit, used only for accounting purposes in the Hapsburg period.

REAL-OF-EIGHT, RIAL-OF-EIGHT Spanish silver peso officially valued at Goa at 400 réis, or the equivalent of one cruzado. The plural (reales, reals, or rials) commonly used to denote simply imported silver coinage.

SÃO TOMÉ A Portuguese gold coin minted at Goa, with a fixed value of 360 réis.

SOLDADO A Portuguese soldier; Portuguese or Eurasian bachelor of military age, in the State of India.

TANGA An Indo-Portuguese silver coin valued at 60 réis.

TERÇO A standing infantry regiment with a full strength of about 3,000 men; the Portuguese equivalent of the Spanish *tercio*.

VEDOR DA FAZENDA Treasurer; comptroller.

VEDOR DA FAZENDA GERAL Comptroller-general.

VINTÉM A penny; 20 réis.

XERAFIM (pl. XERAFINS) Standard silver coin of Portuguese India, valued at 300 réis.

Bibliography

Manuscript Sources

Portugal

Arquivo Histórico Ultramarino, Lisbon
  Codices 218, 281-283, 432, 476, 1164
  Documentos Soltos Relativos a Índia, Caixas 9-10
Arquivo Nacional da Tôrre do Tombo, Lisbon
  Codex 319
  Inquisição de Lisboa, Processos N4474, N7703
  Livros das Monções 12-47
Biblioteca da Academia das Ciências, Lisbon
  Codex 312A
  Ms Vermelho 559
Biblioteca da Ajuda, Lisbon
  Codices 46-xiii-30, 49-v-8, 49-x-28, 49-xii-28, 49-xii-38, 49-xiii-12, 50-v-37, 51-vi-2, 51-vii-12, 51-vii-13, 51-vii-30, 51-vii-32, 51-x-2
Biblioteca Nacional, Lisbon
  Fundo Geral, Codices 939, 1783, 7640; Caixa 210, no. 115
  Collecção Pombalina, Codices 313, 490
Biblioteca Pública e Arquivo Distrital, Évora
  Codices cv/2-7, cxv/1-5, cxv/1-21, cxvi/1-18, cxvi/2-3, cxvi/1-39

Spain

Archivo General de Simancas
  Estado, legajos 435-437
  Secretaries Provinciales, codex 1571

England

British Library, London
  Additional MS. 20,902
Library of King's College, University of London
  MS 14 Codex Lynch

United States

Houghton Library, Harvard University
  MS Portug 4794F
Newberry Library, Chicago
  Greenlee Collection, MS.124

Brazil
Biblioteca Nacional, Rio de Janeiro
Codices 1-13/2-1, 1-13/2-6, 2/2-19, 10/3-27
Codices Pernambuco 1/2-35, 1/2-36
Instituto Histórico e Geográfico Brasileiro, Rio de Janeiro
Lata 73, doc. 23
India
Archive of the Indies, Panjim
Codices 415, 1161, 1662, 2358, 2608, 7738, 7745, 7786, 7846, 10397
Libros das Monções 13A, 13B, 14, 15, 16A, 16B, 17, 18, 19A, 19B

Printed Sources

Andrade e Silva, José Justino de, comp., *Collecção chronológica da legislação portugueza.* 10 vols., Lisbon, 1854-1857.

Baldaeus, Philippus, "A True and Exact Description of the Most Celebrated East-India Coasts of Malabar and Coromandel as also of the Isle of Ceylon . . . ," in Awnsham Churchill and John Churchill, comps. *A Collection of Voyages and Travels* . . . . 4 vols., London, 1672, III, 561-901.

Barbosa, Duarte, *The Book of Duarte Barbosa, an Account of the Countries Bordering on the Indian Ocean and their Inhabitants, Written by Duarte Barbosa and Completed About the Year 1518 A.D.* Trans. and ed. Mansel Longworth Dames. 2 vols., Hakluyt Society, 2nd series, no. 44, London, 1918-1921.

Biker, Júlio Firmino Júdice, *Collecção de tratados e concertos de pazes que o Estado da India Portugueza fez com os reis e senhores com quem teve relações nas partes da Asia e Africa Oriental desde o principio da conquista até ao fim do seculo XVIII.* 14 vols., Lisbon, 1881-1887.

Birdwood, George, and Foster, William, eds., *The First Letter-Book of the East Company 1600-1619.* London, 1893.

Bocarro, António, *Decada 13 da historia da India.* Ed. Rodrigo José de Lima Felner. Lisbon, 1876.

———, "Livro das plantas de todas as fortalezas, cidades e povações do Estado da Índia Oriental," in A.B. de Bragança Pereira, ed., *Arquivo Português Oriental,* bk.4, vol.2, pt. 1-3.

Boletim da Filmoteca Ultramarina Portuguesa. Ed. António da Silva

Rego. Lisbon, Centro de Estudos Históricos Ultramarinos, 1954-.

Botelho, Simão, "Tombo do Estado da India," in Rodrigo José de Lima Felner, ed., *Subsídios para a historia da India Portugueza.* Lisbon, 1868.

Boxer, C. R., *The Tragic History of the Sea, 1589-1622.* Hakluyt Society, 2nd series, no.112, Cambridge, 1959.

Bragança Pereira, A. B. de, *Arquivo português oriental.* 11 vols., Bastorá-Goa, Tipografia Rangel, 1936-1940.

Bulhão Pato, Raymundo António de, ed., *Documentos remettidos da India, ou Livros das monções.* 5 vols., Lisbon, Academia das Sciencias, 1880-1935.

Cabreira, Joseph de, *Naufragio da Noa Nossa Senhora de Bethlem. Feito na terra do Natal no Cabo de Boa Esperança.* Lisbon, 1636.

Castanheda, Fernão Lopes de, *Historia do descobrimento e conquista da India pelos portugueses* [1551-1561]. 3rd ed. 9 vols. in 4, Coimbra, Imprensa da Universidade, 1924-1933.

Conceição, Nuno da, "Relação da viagem e successo que teve a nao capitania Nossa Senhora do Bom Despacho . . . ," in Bernardo Gomes de Brito, comp., *História tragico-maritima* [1735-1736], ed. Gabriel Pereira, Lisbon, 1904-1909, IX, 107-128.

Costa, Francisco da, "Relatorio sobre o trato da pimenta feito por Francisco da Costa, escrivão da feitoria de [Cochim]," in *Documentação Ultramarina Portuguesa,* ed. António da Silva Rego, III, 295-379.

Couto, Diogo do, *Da Asia de Diogo do Couto* [1602-1736]. 12 vols., Lisbon, 1777-1788.

Cunha Rivara, J. H. da, ed., *Archivo portuguez-oriental.* 6 vols. in 9, Nova Goa, 1857-1876.

Falcão, Luiz de Figueiredo, *Livro em que se contém toda a fazenda e real patrimonio dos reinos de Portugal, India e ilhas adjacentes e outras particularidades* [completed 1612, or soon after]. Lisbon, 1859.

Faria e Sousa, Manuel de, *Ásia portuguesa* [1663-1675]. 6 vols. Oporto, Livraria Civilização-Editora, 1947.

Foster, William, *The English Factories in India, 1618-1669.* 13 vols., Oxford, Clarendon Press, 1906-1927.

Freire de Oliveira, Eduardo de, *Elementos para a história do município de Lisboa.* 17 vols., Lisbon, 1887-1898.

Gentil da Silva, J., ed., *Alguns elementos para a história do comercio da Índia de Portugal existentes na Biblioteca Nacional de Madrid. Anais,* vol. 5, no.2, Junta das Missões Geográficas e de Investigações Coloniães, 1950.

Gomes de Brito, Bernardo, comp. *História trágico-marítima* [1735-1736]. Ed. Gabriel Pereira. 12 vols., Lisbon, 1904-1909.

Gomes Solis, Duarte, *Alegación en favor de la Compañia de la Índia Oriental* [1628]. Ed. Moses Bensabet Amzalak, Lisbon, Editorial Império, 1955.

Herbert, Thomas, *Some Yeares Travels into Divers Parts of Africa and Asia the Great Especially Describing the famous Empires of Persia and Industant As also Divers other Kingdoms in the Orientall Indies, and Iles Adjacent.* 3rd edition, London, 1677.

Linhares, Conde de, *Diário do terceiro conde de Linhares, vice-rei da Índia.* Lisbon, Biblioteca Nacional, 1937-1943.

Linschoten, Jan Huygen van, *Itinerario voyage ofte schipvaert van Jan Huygen van Linschoten naer cost ofte Portugaels Indien 1579-1592.* Ed. H. Kern, 3 vols., S-Gravenhage, Martinus Nijhoff, 1955-1957.

———, *The Voyage of John Huyghen van Linschoten to the East Indies.* Ed. Arthur Coke Burnell and P. A. Tiele. 2 vols., Hakluyt Society, 1st series, nos. 70-71, London, 1884.

Mandelslo, Jean Albert de [Johann Albrecht von Mandeslo] *Voyages celebres e remarquables, faits de Perse aux Indes Orientales* [1658]. Trans. A. de Wicquefort. 2 vols., Amsterdam, 1727.

Mendes da Luz, Francisco Paulo, ed., *Livro das cidades e fortalezas que a coroa de Portugal tem nas partes da India, e das capitanias, e mais cargos que nelas ha, e da importancia delles.* Facsimile, Lisbon, Centro de Estudos Históricos Ultramarinos, 1960.

———, *Regimento da Caza da India. Manuscrito do século XVII existente no Arquivo Geral de Simancas. Anais,* Junta das Missões Geográficas e de Investigações Coloniães, vol. 6, no. 2, 1951.

Mocquet, Jean, *Travels and Voyages into Africa, Asia, and America, the East and West Indies, Syria, Jerusalem and the Holy-Land* [1617]. Trans. Nathaniel Pullen. London, 1696.

Mundy, Peter, *The Travels of Peter Mundy in Europe and Asia, 1608-1667.* Ed. Sir Richard Carnac Temple and Lavinia Mary Anstey. 5 vols. in 6, Hakluyt Society, 2nd series, nos. 17, 35, 45, 46, 55, 78, London, 1907-1936.

Orta, Garcia d', *Colloquies on the Simples and Drugs of India.* Trans. and introd, Sir Clements Markham. London, Henry Sotheran and

Co., 1913.
———, *Coloquios dos simples e drogas e cousas medecinais da India* . . . , [1563]. Lisbon, Academia das Ciências, 1963.
Pissurlencar, Panduronga S. S., ed., *Assentos do Conselho do Estado 1618-1750*. 5 vols., Bastorá-Goa, Tipografia Rangel, 1953-1957.
———, *Regimentos das fortalezas da Índia*. Bastorá-Goa, Tipografia Rangel, 1951.
Pyrard de Laval, François, *The Voyage of François Pyrard of Laval to the East Indies, the Maldives, the Moluccas and Brazil* [1611]. Ed. and trans. Albert Gray and H. C. P. Bell. 2 vols. in 3, Hakluyt Society, 1st series, nos. 76, 77, 80, London, 1887-1890.
Ribeiro, João, *Ribeiro's History of Ceilão with a summary of De Barros, De Couto, Antonio Bocarro, and the Documentos Remettidos, with the Parangi Hatane and Konstantinu Hatane*. Trans. and ed. P. E. Pieris. Colombo, Colombo Apothecaries Co., 1909.
Ribeiro, Luciano, *Registo da Casa da Índia*. 2 vols., Lisbon, Agência Geral do Ultramar, 1954.
Santa Maria, Agostinho de, *História da fundação do real convento de Santa Monica*. Lisbon, 1699.
Silva Rego, António da, ed., *Documentação para a história das missões do padroado português do Oriente. Índia*. 12 vols. Lisbon, Agência Geral do Ultramar, 1947-1958.
———, *Documentação ultramarina portuguesa*. Lisbon, Centro de Estudos Históricos Ultramarinos, 1960-.
Tavernier, Jean-Baptiste, *Travels in India* [1676]. Trans. and ed. V. Ball. 2 vols. London and New York, 1889.
Theal, George McCall, *Records of South-Eastern Africa*. 9 vols., Cape Town, 1898-1903.
Trindade, Paulo da, *Conquista espiritual do Oriente*. Ed. F. Felix Lopes, 3 vols., Lisbon, Centro de Estudos Históricos Ultramarinos, 1962-1967.
Valle, Pietro della, *The Travels of Pietro della Valle in India* [1663]. Trans. G. Havers, 1664. Ed. Edward Grey. 2 vols., Hakluyt Society, 1st series, nos. 84-85, London, 1892.

Selected Secondary Works

Abeyasinghe, Tikiri, *Portuguese Rule in Ceylon 1594-1612*. Colombo, Lake House, 1966.
Axelson, Eric, *Portuguese in South-East Africa 1600-1700*. Johannes-

burg, Witwatersrand U. P., 1960.

Azevedo, João Lúcio de, *Épocas de Portugal económico*. Lisbon, Livraria Clássica Editora, 1929.

———, *História dos Christãos Novos portugueses*. Lisbon, Livraria Clássica Editora, 1922.

Baião, António; Cidade, Hernani; and Múrias, Manuel, eds., *História da expansão portuguesa no mundo*. 3 vols., Lisbon, Editorial Ática, 1937-1940.

Bharucha, R. M., "Anglo-Portuguese Relations (1600-1665)." Ph.D. dissertation, University of Bombay, 1946.

Botelho de Sousa, Alfredo, *Subsídios para a história militar marítima da India, 1585-1669*. 4 vols., Lisbon, Imprensa da Armada, 1930-1956.

Boxer, C. R., *The Dutch in Brazil 1624-1654*. Oxford, Clarendon Press, 1957.

———, *The Dutch Seaborne Empire 1600-1800*. London, Hutchinson, 1965.

———, *Fidalgos in the Far East, 1550-1770*. The Hague, Martinus Nijhoff, 1948.

———, *Francisco Vieira de Figueiredo: A Portuguese Merchant-Adventurer in South-East Asia, 1624-1667*. The Hague, Martinus Nijhoff, 1967.

———, "The General of the Galleons and the Anglo-Portuguese Truce Celebrated at Goa in January 1635," in *Ethnos* 1 (1935), 27-33.

———, "José Pinto Pereira, Vedor da Fazenda Geral da Índia" in *Anais da Academia Portuguesa da História,* 7 (1942), 79-80.

———, "On a Portuguese Carrack's Bill of Lading in 1625," in *Biblos,* 14 (1938), 193-222. Also in *Petrus Nonius,* vol.2, fasc.3 (1939), pp.176-200.

———, "The Naval and Colonial Papers of Dom Antonio de Ataide," in *Harvard Library Bulletin,* 5 (1951), 24-50.

———, "Padre Antonio Vieira S. J., and the Institution of the Brazil Company in 1649," *Hispanic American Historical Review,* 29 (1949), 474-497.

———, "Portuguese and Spanish Projects for the Conquest of Southeast Asia," in *Journal of Asian History,* 3 (1969), 118-136.

———, *The Portuguese Seaborne Empire 1415-1825*. London, Hutchinson, 1969.

———, *Portuguese Society in the Tropics: The Municipal Councils of Goa, Macao, Bahia, and Luanda, 1510-1800*. Madison and Milwaukee, University of Wisconsin Press, 1965.

———, *Race Relations in the Portuguese Colonial Empire 1415-1825*. Oxford, Clarendon Press, 1963.

———, *Salvador da Sá and the Struggle for Brazil and Angola 1602-1686*. London, Athlone Press, 1952.

———, "Three historians of Portuguese Asia (Barros, Couto, Bocarro)," in *Boletim do Instituto Português de Hongkong,* 1 (1948), 15-44.

———, and Azevedo, Carlos de, *Fort Jesus and the Portuguese in Mombasa 1593-1729*. London, Hollis and Carter, 1960.

Buchanan, Francis A., *A Journey from Madras through the Countries of Mysore, Canara and Malabar*. 3 vols., London, 1807.

Burkill, I. H., *A Dictionary of the Economic Products of the Malay Peninsula*. 2 vols., London, Crown Agents for the Colonies, 1935.

Burton, Richard, *Goa and the Blue Mountains: or, Six Months of Sick Leave*. London, 1851.

Caetano de Sousa, António, *História genealógica da casa real portuguesa*. 12 pts., Coimbra, Atlântida Livraria Editora, 1946-1951.

Cambridge History of India. Ed. Sir Richard Burn. 6 vols., Cambridge University Press, 1922-1937.

Carvalho, Tito Augusto de, *As companhias portugueses de colonização*. Lisbon, Imprensa Nacional, 1902.

Chaudhuri, K. N., *The English East India Company. The Study of an Early Joint-Stock Company 1600-1640*. London, Frank Cass, 1965.

Chaunu, Huguette, and Chaunu, Pierre, *Séville et l'Atlantique 1504-1650*. 8 vols., Paris, Librarie Armand Colin, 1955-1959.

Codrington, H. W., *A Short History of Ceylon*. London, Macmillan, 1947.

Cousens, Henry, *Bijapur and Its Architectural Remains with an Historical Outline of the Adil Shahi Dynasty*. Bombay, Government Central Press, 1916.

Cunha Rivara, J. H. da, "A Companhia do Comercio," in *O Chronista de Tissuary,* nos. 17-20 (May-August 1867).

———, "O Idalxa 1629-1633," in *O Chronista de Tissuary,* nos.2-5 (February-May 1866).

Danvers, F. C., *The Portuguese in India. Being a History of the Rise and Decline of their Eastern Empire*. 2 vols., London, 1894.

Das Gupta, A., *Malabar in Asian Trade 1740-1800*. Cambridge University Press, 1967.

Dicionário de história de Portugal. Ed. Joel Serrão. 3 vols., Lisbon, 1965-1968.

Domingues Ortiz, Antonio, *Politica y hacienda de Felipe IV*. Madrid, Editorial de Derecho Financiero, 1960.

Duffy, James, *Portuguese Africa*. Cambridge, Mass., Harvard University Press, 1959.

———, *Shipwreck and Empire. Portuguese Maritime Disasters in a Century of Decline*. Cambridge, Mass., Harvard University Press, 1955.

Fonseca, José Nicolau de, *An Historical and Archeological Sketch of the City of Goa*. Bombay, 1878.

Foster, William, *England's Quest for Eastern Trade*. London, A. and C. Black, 1933.

Gazetteer of the Bombay Presidency. Ed. Sir James Macnabb Campbell. 27 vols. in 36, Bombay, Government Central Press, 1877-1904.

Gentil da Silva, J., *Stratégie des affaires à Lisbonne entre 1595 et 1607. Lettres marchandes des Rodrigues d'Évora et Veiga*. Paris, SEVPEN, 1956.

Glamann, Kristof, *Dutch-Asiatic Trade 1620-1740*. Copenhagen, Danish Science Press, and The Hague, Martinus Nijhoff, 1958.

Hamilton, Walter, *The East India Gazetteer* . . . . 2 vols., London, 1866.

Heras, H., "The Expansion Wars of Venkatapa Nayaka of Ikeri," in *Proceedings of the Indian Historical Records Commission* (1929), pp.106-124.

Hussey, Roland D., "Antecedents of the Spanish Monopolistic Overseas Trading Companies, (1624-1728)," in *Hispanic American Historical Review,* vol.9, no.2 (August 1929), pp.1-30.

———, *The Caracas Company*. Cambridge, Mass., Harvard University Press, 1934.

Imperial Gazetteer of India. Ed. Sir Herbert H. Risley, W. S. Meyer, R. Burn and J. S. Cotton. Revised edition, 26 vols., Oxford, Clarendon Press, 1907-1909.

Joshi, P. M., "The Portuguese on the Deccan (Konkan) Coast: Sixteenth and Seventeenth Centuries," in *Journal of Indian History,* 46, (1968), 65-88.

Kieniewicz, Jan, "The Portuguese Factory and Trade in Pepper in

Malabar during the Sixteenth Century," in *Indian Economic and Social History Review,* 6 (1969), 61-84.

Kellenbenz, Hermann, "Autour de 1600: le commerce du poivre des Fuggers et le marché international du poivre," in *Annales: Économies. Societés. Civilisations,* 11 (1956), 1-28.

Lach, Donald F., *Asia in the Making of Europe.* 2 pts., Chicago and London, U. of Chicago Press, 1965.

Lavanha, João Baptista, *Viage de la catholica Real Magestad del Rei D. Filipe. III. N. S. al reino de Portugal.* Madrid, 1622.

Livermore, H. V., ed., *Portugal and Brazil.* Oxford, Clarendon Press, 1963.

Lopes, David, trans. and ed., *Historia dos Portugueses no Malabar por Zinadim.* Lisbon, 1898.

Macpherson, David, *The History of the European Commerce with India.* London, 1812.

Magalhães, José Calvet de, "Duarte Gomes Solis," *Studia,* no. 19 (December 1966), pp.119-171.

Magalhães Godinho, Vitorino, "Les finances de l'état portugais des Indes Orientales. 1517-1635," Ph.D. thesis, La Sorbonne, Paris, 1958.

———, *Os descobrimentos e a economia mundial.* 2 vols., Lisbon, Editora Arcádia, 1963-1965. See also shortened edition in French: *L'Économie de l'empire portugais aux XVe et XVIe siècles,* SEVPEN, Paris, 1969.

Mauro, Frédéric, *Le Portugal et l'Atlantique au XVIIe siècle 1570-1670.* Paris, SEVPEN, 1960.

Meilink-Roelofsz, M. A. P., *Asian Trade and European Influence in the Indonesian Archipelago between 1500 and about 1630.* The Hague, Martinus Nijhoff, 1962.

Melo, Francisco Manuel de, *Epanáforas de vária história portuguesa* [1660]. Ed. Edgar Prestage. 3rd revised edition, Coimbra, Imprensa da Universidade, 1931.

Melo de Matos, Gastão de, *Notícias do têrço da armada real (1618-1707).* Lisbon, Imprensa da Armada, 1932.

Mendes da Luz, Francisco Paulo, *O Conselho da Índia.* Lisbon, Agência Geral do Ultramar, 1952.

Menon, P. Shungoonny, *The History of Travancore.* Madras, 1878.

Moreland, W. H., *From Akbar to Aurangzeb. A Study in Indian Economic History.* London, Macmillan, 1923.

———, *India at the Death of Akbar*. London, Macmillan, 1920.

Nobreza de Portugal; bibliografia, biografia, cronologia, filatelia, genealogia, heráldica, história, nobiliarquia, numismática. 3 vols., Lisbon, Editorial Enciclopédia, 1960-1961.

Oliveira Marques, A. H. de, *História de Portugal*. Lisbon, Edições Ágora, 1973.

Parry, J. H., *The Age of Reconnaissance. Discovery, Exploration and Settlement 1450 to 1650*. London, Weidenfeld and Nicolson, 1963.

Pearson, Michael N., "Indigenous Dominance in a Colonial Economy. The Goa *rendas,* 1600-1670," in *Mare Luso-Indicum,* 3 (1972), 61-73.

———, *Merchants and Rulers in Gujarat. The Response to the Portuguese in the Sixteenth Century*. Berkeley and Los Angeles, University of California Press, 1976.

———, "Wealth and Power: Indian Groups in the Portuguese Indian Economy," in *South Asia,* no.3 (August 1973), pp.36-44.

Penrose, Boies, *Goa—Rainha do Oriente; Goa—Queen of the East*. Lisbon, Comissão Ultramarina, 1960.

Pieris, P. E., *Ceylon and the Portuguese 1505-1658*. Tellipalai, American Ceylon Missions Press, 1920.

Poonen, T. I., *A Survey of the Rise of the Dutch Power in Malabar 1603-1678*. Trichinopoly, St. Joseph's Industrial School Press, 1948.

Prestage, Edgar, ed., *Chapters in Anglo-Portuguese Relations*. Watford, Voss and Michaels, 1935.

Purseglove, J. W., *Tropical Crops. Dicotyledons*. 2 vols. London, Longmans, 1968.

Rebello da Silva, Luíz Augusto, *História de Portugal nos séculos XVII e XVIII*. 5 vols., Lisbon, 1860-1871.

Rice, Benjamin Lewis, *Mysore and Coorg from the Inscriptions*. London, Archibald Constable and Co., 1909.

Ródenas Vilar, Rafael, "Un gran proyecto anti-holandés en tiempo de Felipe IV," in *Hispania,* 22 (1962), 542-558.

———, *La política Europea de España durante la guerra de Treinte Años (1624-1630)*. Madrid, Consejo Superior de Investigaciones Científicas, 1967.

Sewell, Robert, *A Forgotten Empire (Vijayanagar). A Contribution to the History of India*. London, Swan Sonnenschein and Co., 1900.

Silva, C. R. de, "The Portuguese East India Company 1628-1633, *Luso-Brazilian Review,* 11 (1974), 152-205.

———, *The Portuguese in Ceylon 1617-1638*. Colombo, H. W. Cave and Co., 1972.

———, "Trade in Ceylon Cinnamon in the Sixteenth Century," in *Ceylon Journal of Historical and Social Studies,* n. s., vol.3, no.2 (1973), pp.14-27.

Silva Rego, António da, *História das missões do padroado português do Oriente. India I (1500-1542)*. Lisbon, Agência Geral das Colonias, 1949.

———, *O padroado português do Oriente. Esboço histórico*. Lisbon, Agência Geral das Colonias, 1940.

Silveira, Luís, *Ensaio de iconagrafia das cidades portugueses do ultramar*. 4 vols., Lisbon, Ministério do Ultramar, 1952-1955.

Smith, David Grant, "Old Christian Merchants and the Foundation of the Brazil Company, 1649," in *Hispanic American Historical Review,* vol.54, no.2 (May 1974), pp.233-259.

Spate, O. H. K., and Learmonth, A. T. A., *India and Pakistan. A General and Regional Geography*. London, Methuen, 1967.

Steensgaard, Niels, *Carracks, Caravans and Companies: The Structural Crisis in the European-Asian Trade in the Early 17th Century*. Copenhagen, Scandinavian Institute of Asian Studies, 1973.

Tratado de todos os vice-reis e governadores da Índia. Lisbon, Editorial Enciclopédia, 1962.

Vasconcelos, Frazão de, "A marinha da coroa de Portugal no tempo dos Felipes," in *Congresso do Mundo Português: Publicações,* 6 (1940), 251-264.

Vidago, João, "Anotações a uma bibliografia da 'Carreira da Índia,' " in *Studia,* no.18 (August 1966), pp.209-241.

Vlekke, Bernard H. M., *Nusantara, A History of the East Indian Archipelago*. Cambridge, Mass., Harvard University Press, 1943.

Watt, G., *A Dictionary of the Economic Products of India*. 6 vols., London and Calcutta, 1889-1896.

Wigram, Herbert, and Moore, Lewis, *Malabar Law and Custom*. Madras, Higginbotham and Co., 1900.

Winius, George Davison, *The Fatal History of Portuguese Ceylon. Transition to Dutch Rule*. Cambridge, Mass., Harvard University Press, 1971.

Index

Harvard Historical Studies

Out of Print Titles are Omitted

33. *Lewis George Vander Velde.* The Presbyterian Churches and the Federal Union, 1861-1869. 1932.
35. *Donald C. McKay.* The National Workshops: A Study in the French Revolution of 1848. 1933.
38. *Dwight Erwin Lee.* Great Britain and the Cyprus Convention Policy of 1878. 1934.
48. *J. H. Hexter.* The Reign of King Pym. 1941.
58. *Charles C. Gillispie.* Genesis and Geology: A Study in the Relations of Scientific Thought, Natural Theology, and Social Opinion in Great Britain, 1790-1850. 1951
60. *Robert G. L. Waite.* Vanguard of Nazism: The Free Corps Movement in Post-war Germany, 1918-1923. 1952.

62, 63. *John King Fairbank.* Trade and Diplomacy on the China Coast: The Opening of the Treaty Ports, 1842-1854. One-volume edition. 1953.

64. *Franklin L. Ford.* Robe and Sword: The Regrouping of the French Aristocracy after Louis XIV. 1953.
66. *Wallace Evan Davies.* Patriotism on Parade: The Story of Veterans' and Hereditary Organizations in America, 1783-1900. 1955.
67. *Harold Schwartz.* Samuel Gridley Howe: Social Reformer, 1801-1876. 1956.
69. *Stanley J. Stein.* Vassouras: A Brazilian Coffee Country, 1850-1900. 1957.
72. *John B. Blake.* Public Health in the Town of Boston, 1630-1822. 1959.
74. *Alexander Sedgwick.* The Ralliement in French Politics, 1890-1898. 1965.
75. *E. Ann Pottinger.* Napoleon III and the German Crisis, 1865-1866. 1966.
76. *Walter Goffart.* The Le Mans Forgeries: A Chapter from the History of Church Property in the Ninth Century. 1966.
77. *Daniel P. Resnick.* The White Terror and the Political Reaction after Waterloo, 1966.
78. *Giles Constable.* The Letters of Peter the Venerable. 1967.
79. *Lloyd E. Eastman.* Throne and Mandarins: China's Search for a Policy during the Sino-French Controversy, 1880-1885. 1967.
80. *Allen J. Matusow.* Farm Policies and Politics in the Truman Years. 1967.
81. *Philip Charles Farwell Bankwitz.* Maxime Weygand and Civil-Military Relations in Modern France. 1967.
82. *Donald J. Wilcox.* The Development of Florentine Humanist Historiography in the Fifteenth Century. 1969.
83. *John W. Padberg, S.J.* Colleges in Controversy: The Jesuit Schools in France from Revival to Suppression, 1813-1880. 1969.

84. *Marvin Arthur Breslow.* A Mirror of England: English Puritan Views of Foreign Nations, 1618-1640. 1970.
85. *Patrice L. R. Higonnet.* Pont-de-Montvert: Social Structure and Politics in a French Village, 1700-1914. 1971.
86. *Paul G. Halpern.* The Mediterranean Naval Situation, 1908-1914. 1971.
87. *Robert E. Ruigh.* The Parliament of 1624: Politics and Foreign Policy. 1971.
88. *Angeliki E. Laiou.* Constantinople and the Latins: The Foreign Policy of Andronicus, 1282-1328. 1972.
89. *Donald Nugent.* Ecumenism in the Age of the Reformation: The Colloquy of Poissy. 1974.
90. *Robert A. McCaughey.* Josiah Quincy, 1772-1864: The Last Federalist. 1974.
91. *Sherman Kent.* The Election of 1827 in France. 1975.
92. *A. N. Galpern.* The Religions of the People in Sixteenth-Century Champagne. 1976.
93. *Robert G. Keith.* Conquest and Agrarian Change: The Emergence of the Hacienda System on the Peruvian Coast. 1976.
94. *Keith Hitchins.* Orthodoxy and Nationality: Andreiu Şaguna and the Rumanians of Transylvania, 1846-1873. 1977.
95. *A. R. Disney.* Twilight of the Pepper Empire: Portuguese Trade in Southwest India in the Early Seventeenth Century. 1978.

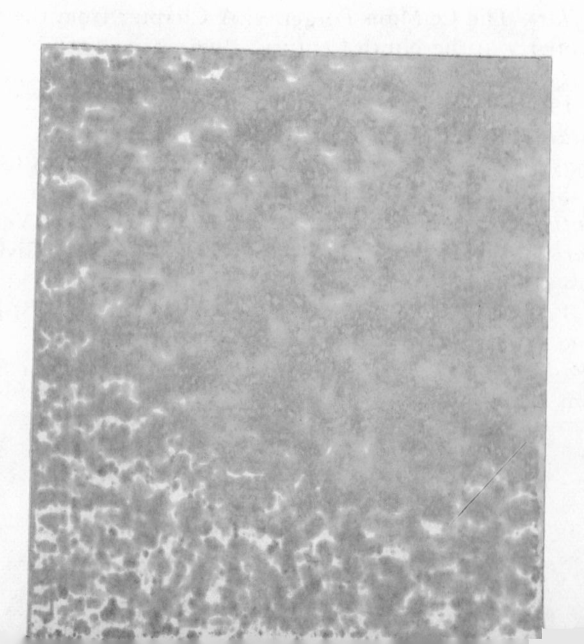